Habib Fekih, who is Tunisian, is an aerospace electronics engineer who has worked in the fields of both airlines and aircraft manufacturers. He has held several executive positions, which allowed him to become a privileged witness of the aviation industry. Habib participated in the commercial success of Airbus and in several projects which impacted the industry, and he has built relationships with many famous personalities around the world. His passions for both aviation and history allow him to take us on a journey through the world of aviation for eight decades.

I dedicate this book to my wife, Ibtissem, and my son, Karim, who supported me for more than 40 years and stood by my side in all my ups and downs. I also dedicate this book to all those colleagues, and dare I say friends, with whom I have worked for so many years and without whom none of what was achieved could have been achieved—particularly to those who have left us (RIP). All of them contributed to making aviation what it is today and made me enjoy working in it. I am also grateful to all the members of the various teams I had to work with for their contribution to our numerous joint successes. All these colleagues, friends, and team members are so numerous that I could not mention them all in this book. Yes, I don't forget them, nor do I forget their contribution to our joint successes. Thank you to you all! I also dedicate this book to all the youngsters who have dreams and wonder if they can make their dreams come true. This book should convince them that, with tenacity, dedication, and hard work, no matter their origins, it is possible. Just never give up!

Habib Fekih

Fly High, Fly Low, Fly Fast, Fly Slow

Austin Macauley Publishers

LONDON • CAMBRIDGE • NEW YORK • SHARJAH

Copyright © Habib Fekih 2025

The right of Habib Fekih to be identified as the author of this work has been asserted by the author in accordance with sections 77 and 78 of the Copyright, Designs and Patents Act 1988.

All rights reserved. No part of this publication may be reproduced, stored in a retrieval system, or transmitted in any form or by any means, electronic, mechanical, photocopying, recording, or otherwise, without the prior permission of the publishers.

Any person who commits any unauthorised act in relation to this publication may be liable to criminal prosecution and civil claims for damages.

All of the events in this memoir are true to the best of author's memory. The views expressed in this memoir are solely those of the author.

A CIP catalogue record for this title is available from the British Library.

ISBN 9781035873357 (Paperback)
ISBN 9781035873364 (Hardback)
ISBN 9781035873388 (ePub e-book)
ISBN 9781035873371 (Audiobook)

www.austinmacauley.com

First Published 2025
Austin Macauley Publishers Ltd®
1 Canada Square
Canary Wharf
London
E14 5AA

My sincere thanks to: Sir Tim Clark, President of Emirates Airlines, for his inspiring discussions, his proofreading, and his comments. My colleagues at Tunis Air—Ammar Trabelsi, Hamadi Thamri, and Moncef Ben Dhahbi—for their proofreading and their comments, and in particular Ammar for his computer assistance. My friend and colleague at EADS, Pierre Bayle, for his proofreading and comments. My friend and colleague at Airbus, Barbara Kracht, for her multiple proofreadings, her suggestions, comments, and corrections. My grandson, Kaïs Fekih, for his inspiring and beautiful drawing used on the front cover of this book.

Table of Contents

Foreword (Before Take-Off) (Hit the Road Jack-Ray Charles) — 13

1. Awakening (Awaken—Yes) — 16

   1.1 Bip-Bip in the Sky (Johnny B. Goode-Chuck Berry) — 18

   1.2 A Dog in Space (Breathless-Jerry Lee Lewis) — 20

   1.3 Death from the Sky (The End-The Doors) — 22

   1.4 The Monster (Steppenwolf) — 23

   1.5 Aircraft Noise (Beautiful Noise-Neil Diamond) — 25

   1.6 A Man in Space (Rocketman—Elton John) — 27

2. A Teenager Passion (Teenage Rampage—Sweet) — 30

   2.1 Living Next to an Airport (Airport—The Motors) — 30

   2.2 The Beatles and the Pan Am 707 (Hold Your Hand-The Beatles) — 34

   2.3 First Close Sight of an Aircraft (In the Army Now-Status Quo) — 35

   2.4 Walking on the Moon (The Police) — 37

   2.5 First Air Travel with a Caravelle (Give Me a Ticket for an Airplane—The Box Tops) — 41

3. Student's Dreams and Challenges (Ô Toulouse—Claude Nougaro) — 43

   3.1 A Supersonic Encounter—Concorde (Speed of Sound—Coldplay) — 43

   3.2 Loss of a Close Friend in an Aircraft Accident (Stairway to Heaven—Led Zeppelin) — 51

   3.3 Paris, The City of Light (Paris s'éveille—Jacques Dutronc) — 51

    3.4 Engineering Dream$ (Sweet Dreams—Eurythmics)   53

    3.5 Information Technology is the Future (Communication Breakdown—Led Zeppelin)   54

    3.6 The World of Quantum Mechanics (Atom Heart Mother—Pink Floyd)   55

    3.7 Change of Direction (Crossroads—Eric Clapton)   57

    3.8 Another Aerospace Horizon (Interstellar Overdrive—Pink Floyd)   65

    3.9 Get Back to Where You Once Belonged (The Beatles)   68

4. Learning the Aviation Job in Tunisia (Night in Tunisia-Dizzy Gillispie)   **70**

    4.1 Landing at Tunis Air (4&20 Years Ago-Stephen Stills)   70

    4.2 Airline Daily Life (Sultan of Swings-Dire Straits)   72

    4.3 The Airbus A300-B4 FFCC (The Two-Man Crew Saga) (Satisfaction-The Rolling Stones)   83

    4.4 Tunis Air at the Frontline (Front Line-Stevie Wonder)   96

    4.5 Developing and Reforming the Airline (Bohemian Rhapsody-Queen)   100

    4.6 Managing a Stupid Survival Crisis (Under Pressure—Queen)   107

5. Practicing the Aviation Job Worldwide (On the Road Again-Canned Heat)   **115**

    5.1 Airbus GIE (A Group of Pioneers—1986–1998) (Simply the Best-T.Turner)   116

    5.2 Airbus Corporation (More Professional, but Red Tapes) (Take a chance on me-ABBA)   176

    5.3 Airbus Group (Major Tom) (Space Oddity-David Bowie)   197

    5.4 The Englishman (Sir Tim Clark) (English Man in New York-Sting)   208

    5.5 The Arab Spring (Won't Get Fooled Again- The WHO)   239

6. Citizen of the World (We are the World- USA for Africa)   **244**

    6.1 The World of Aviation (Jump-Van Halen)   244

6.2 Aviation and Politics (Political World-Bob Dylan)	*257*
6.3 Aviation Leaders (Airlines' Leaders) (The Boss-James Brown)	*260*
6.4 World Leaders (World Leaders-Rem)	*266*
6.5 Where is Home (Sweet Home Alabama-Lynyrd Skynyrd)	*267*
6.6 The Global Village (In the Name of Love-U2)	*269*
7. Lessons Learnt (I Believe I Can Fly)	**272**
7.1 Aviation is a Real WWW (Lift Me Up-Moby)	*272*
7.2 Keep on Flying (Keep on Running-Spencer Davis Group)	*273*
8. The Future (In the Year 2525-Zager & Evans)	**275**
Conclusion (Before Landing) (Rien De Rien-Edith Piaf)	**278**
Playlist	**280**

Foreword
(Before Take-Off) (Hit the Road Jack-Ray Charles)

Some friends and ex-colleagues suggested to me, after my retirement, to write something about my personal aerospace experience. They believe that certain aspects could be of interest to the public and to the youngsters.

Due to my education and my family values, which promote humility and discretion, I was reluctant to speak about myself. In addition, I am still deeply convinced that I am not a public figure whose life, deeds, and words may be interesting for the public.

A biography must not be a simple, silly exercise for the sole sake of satisfying the ego of the concerned person.

I am a history buff and love putting history bricks on top of each other to better understand the sequence of events, the reasons behind them, and their consequences on the lives of the people, the countries, and the world.

When we speak about history, we imply a certain chronology and a timeline that can be used as a reference. In the case of a biography or an autobiography, the timeline is simply the life of the person (Lifeline).

Consequently, time becomes a sort of train we get on at our birth and we descend from at our death. It follows its own track, and we have no control over it, besides choosing our seat (if we can), the people we interact with (again, if we can), and the things we do or words we say.

One day, I was visiting Wadi Rum in Jordan, and an old Bedouin, who saw me complaining about the long delay of our bus, told me:

"You can't control time; you can just insert yourself into it."

He was right, but I believe that if we cannot control the course of time, we can, nevertheless, bring changes to things we can control.

I made a quick review of my life and noticed that there are some events of which I was a privileged witness, and which had or even still have an impact on my country, on other countries, and on the world. It happened that these events were, in one way or another, directly or indirectly, linked to aerospace.

Since I spent more than 40 years in this field of activity, I thought that it could be useful to describe these events and shed light on some aspects which are ignored by the public but are, in my opinion, relevant to explain why things happened the way they happened.

I was born in April 1952, in a small village called Mazdour in the Sahel region, in Tunisia, a country where I lived for a large part of my youth.

Mazdour is a village where most of the population is composed of farmers who make their living thanks to olive oil, wheat, barley, melons, watermelons, and prickly pear fruit. The village smells crushed olives and oil in the winter and wheat straws and sweet melon in the summer.

Mazdour bet on education more than a century ago by building one of the first bilingual (Arabic and French) primary schools in the country. These modest farmers did not hesitate to sell some of their olive trees to pay for the education of their children. This led to Mazdour being the birthplace of many people who became celebrities in the country (the first dermatologist, the first PhD in modern Arab literature, the first pilot, one of the first lawyers) and of many high-ranked officers in the army and senior state officials.

I was born into a family with a long military tradition. My father finished his career as chief warrant officer of the Tunisian National Guard (Gendarmerie), of which he was one of the founding members, after serving in the French army for close to 20 years and fighting on many battlefields, from Europe during WWII and subsequently even in Indochina. My great-grandfather was an officer who fought side by side with the French and British armies in Crimea. In each generation, our family had one or two of its members serving in the army, very often as officers.

This military background has tremendously influenced our way of life.

Tunisia was colonised by France and got its independence in 1956. My birthday was three months after the beginning of the armed insurrection against France, on January 18, and 15 days before the first commercial flight of a jet-powered passenger aircraft, the Comet 1, by BOAC.

I consider myself lucky to be a citizen of an independent Tunisia and to live during the new era of civil aviation. I had the chance to be the privileged witness of many major events which transformed the world in many fields, including aerospace. I also had several opportunities to get in close contact with other technologies which became an essential part of our modern life. Last but not least, numerous circumstances linked to my aviation background allowed me to meet prominent politicians and to be present at important political events. I visited dozens of countries, met thousands of people, and came across different cultures, becoming, de facto, a citizen of the world.

In reporting the different events or in speaking about some people, I integrate the surrounding environment with all its relevant components, which influenced my interpretation (social life, political situation, popular music, etc.).

Therefore, references to these environmental components would be quite frequent, and they are surely reflecting my sensitivities. But they have no effect on the veracity of what is reported.

It is important to see the world of aerospace and the world in general from the perspective of the developing countries' citizens, with different references and backgrounds. For decades, aerospace was considered out of reach for the citizens of the Third World, as it was called at the time—*no white, no flight*. Things started changing from the 50s, initially at a slow pace. But thanks to a strong desire from the world's developing countries to have their independence in this field, a real acceleration took place from the mid-60s and led to the integration of aviation in the local economic landscapes, with self-sufficiency in terms of pilots, maintenance, and management. This fundamental transformation did not get the coverage it deserved.

In addition, despite the presence of citizens of the developing world in many key positions within the leading aviation corporations and institutions, we barely hear about their achievements, while their Western colleagues are in the headlines every day, even for non-relevant events.

These are additional reasons which made me decide to write this book.

It happens that I express some personal feelings or some political opinions because they are part of the lived events, while they don't make me the focal point of the story. When you see something beautiful, you can say it is beautiful.

Finally, as music is an important part of my life and as I have always linked some of my favourite songs to periods or events of my life, I kept the connection in this book through dual titles (when and where applicable).

1. Awakening
(Awaken—Yes)

On 23 September 1913, Tunisia, my home country, was amongst the first few countries in the world to experience an aircraft flight, with the arrival of Roland Garros, who gave his name to the famous tennis grand slam tournament in Paris and of whom many people think he was a tennis champion (he played tennis, but he was not a champion).

This was the first air crossing of the Mediterranean Sea, from Fréjus in France to Bizerte in Tunisia, in less than eight hours, with a Blériot aircraft. This great event of French aviation was witnessed by just a few members of the French community in Tunisia while not creating any awareness of aviation among the Tunisian citizens.

Tunisia was out of the burgeoning aviation activity in Europe and in the United States.

That is why I cannot say that my family was acquainted, directly or indirectly, with aviation, besides two major exceptions:

-During World War II, my father and my elder uncle were in the French army, fighting against the Germans, in France, Tunisia, and Belgium. Both, at different locations and periods, were several times attacked by the Luftwaffe. In the case of my father, he became a prisoner of war (POW) on 21 May 1940 in Belgium after his regiment suffered heavy bombardment from the Luftwaffe. He found himself surrounded by German troops for a few days, without ammunition and without food (cf my book *L'Olivier dans la Neige*).

-During that same war, our village, Mazdour, in the region known as the Sahel in Tunisia, which is close to the city of Sousse, witnessed several flights of Allied and German aircraft squadrons heading to this city and its port, with the purpose to carry out bombardments and machine-gunning.

Twice, the attacks took place at less than two miles from our village, at a crossroad, and the villagers could hear the sirens of the Stuka, which were one of the most frightening aspects of the attacks.

There was a striking difference between the Allied attacks, which were very often massive carpet bombing to dislodge the Germans from their fortifications, and the Luftwaffe attacks, which were more selective, focusing on specific targets and, very often far from inhabited areas, at the exception of the Sousse harbour. The number of civilian casualties was always higher following Allied attacks than following German attacks.

There was a strong feeling among the Tunisian citizens that the Allied forces considered them enemies. The pro-German French residents in Tunisia tried to get themselves a new virginity by switching sides and launching strong propaganda against the Tunisian people and their king within the Allied forces and the pro-De Gaulle freedom fighters, like General Juin, by accusing the Tunisians to be pro-German.

This led to the deposition of King Moncef Bey, who was simply nationalistic and defending the interest of his people. The error of the Allied forces was that they trusted French officials who were previously collaborating with the Germans.

They forgot the dozens of thousands of Tunisian soldiers fighting on their side against the Germans, who proved their bravery in many battles in North Africa and who confirmed their loyalty to the Allied forces later on during the battles which took place in Italy, at Monte Cassino and le Belvedere, and on several other frontlines. The first French soldier who crossed the border with Germany was Tunisian, with a special mention of the 4ème RTT (4th Tunisian Riflemen Regiment), which was the most decorated regiment of the whole French army and in which my father had the honour to serve during WWII and during the war of Indochina.

In conclusion, the Tunisian people suffered a double punishment: they were colonised and endured the stupid behaviour of the pro-Nazi French officials, who had attempted to apply Vichy laws—particularly against the Jewish population—but were stopped by the king. At the same time, there was no recognition of the Tunisians' situation by the Allied forces.

Therefore, the only stories I heard about aircraft when I started being interested in my family history were those related to the war, and they were very attractive because they were full of adventures. Some of them were illustrated by specific songs in the Tunisian language, like the famous one Tayar Jana which says, *"The aviator is approaching, may God be with us. You hide left, you hide right, he will always get you at sight."* It continues with words like, *"...Morning and afternoon, blood irrigates the thirsty earth."*

This was a very scary way to get introduced to the world of aviation. Even though I could not grasp the full extent of the stories I was hearing, I got the impression that aircraft were a source of trouble and death. Unfortunately, future events would confirm this impression.

But luckily, a succession of more peaceful (at least at first sight) events gave me the opportunity to build a balanced knowledge about the sky, space, and the different means of travel across the near and distant neighbourhood of the earth.

Childhood is a very interesting period because it is the segment of the life of a human being where discoveries and surprises are a daily routine. Everything is new, and learning is continuous. The bare reality of our parents' daily lives is a source of discoveries.

But if we have the chance to come across really new and unique events which never happened before, there is a sort of amplification of all this learning process, and we go through a sort of quantum jumps, which mark our lives forever. This is not very common, but my generation was blessed by an incredible succession of events which transformed the life of the whole of humanity.

1.1 Bip-Bip in the Sky (Johnny B. Goode-Chuck Berry)

Besides the usual learning to watch the sun with precaution, to watch the different phases of the moon, and to watch the multitude of the stars, I never focused my attention on the sky before I was five and a half years old, more precisely on 6 October 1957.

I was in our village with our maternal grandparents and my brother because my parents were in Tunis for an important appointment with a famous professor of medicine, who was a great specialist in cancer. My mother was diagnosed with a malignant brain tumour in September 1956 and had practically lost her sight.

Therefore, I was not in good shape because I was aware that my mother was seriously ill. But despite this, I was joyful and excited when I dragged one of my cousins and three friends to our paternal grandfather's farm as early as 7 am to locate the first man-made satellite, called *Sputnik 1*, and hear its *Bip-Bip*.

The night before, my maternal grandfather, after listening to the news, told us that the Russians (meaning the Soviet Union) had sent a sort of vehicle into space and made it turn around the Earth while emitting a sound like *Bip-Bip*. Without grasping the full importance of the event, I asked him how we could hear the *Bip-Bip*. He answered that he did not know. This was not helpful.

We were lying on the fresh soil, on the border of a plantation of olive trees, to have a perfect view of the sky with the widest possible angle. There were only some scattered clouds, and most of the sky was clear. The air was pure and fresh, but at an acceptable temperature.

After close to two hours of fruitless observation, we heard the voice of my paternal grandfather, who shouted:

"What are you doing there?"

I jumped in his direction to tell him the truth. After listening to me, he laughed and said:

"You have no chance to hear any noise. Only the Russians could, with special equipment. Now, about seeing Sputnik, it is a question of trajectory and many other parameters."

He then told us:

"If you are interested, come to my office, and I will explain to you what all this means."

My grandfather was comparatively higher educated than people of his generation and had a strong scientific background, despite his job as a regional officer managing state-owned lands.

Since the distance between the farm and the village was rather short, my grandfather led us in a procession to walk towards the village.

While doing so, we heard the bawling of a donkey, and my grandfather reacted by saying:

"Finally, here you have your *Bip-Bip*."

The whole group fell into laughter, and I ran towards my grandfather to hug him and apologise for my stupid idea.

His answer was:

"No, I am proud of you because you showed a great sense of curiosity, and this is good."

After reaching his office, we had a clear explanation about how Earth is a globe, the difference between the sky (the close surrounding envelope of air) and the outer space (which is empty), and how a satellite circles around Earth. All my friends and my cousin were impressed just by the mere fact of being inside this office.

But for me, it was a routine because it was the place where I was coming every day to spend some time with my grandfather while learning Arabic and French. I was already speaking both Tunisian (slightly different from pure Arabic), because it was my mother tongue, and French, because I had lived all my early years in close contact with French children in the barracks of the French Gendarmerie.

For the last year, due to my mother's sickness, I had to come quite often to the village, hence losing the opportunity to go to the normal kindergarten. My grandfather was my teacher—and a very good teacher at that. I learnt both the Arabic and French alphabets, started basic reading and writing, and doing basic calculations.

My maternal grandfather was much more focused on teaching me all about the life of a farmer and everything about olive trees, grains (wheat and barley), melons, and watermelons. This second aspect of my education made me keep my feet on the ground, even if I aspired to fly high.

My two grandfathers and my father had a great impact on my life, my education, and my professional career. They taught me the love of our country, our village, our land, and particularly our olive trees.

They taught me our country and family history, how to work the land, discipline, effort, and perseverance, and, above all, how to set a goal for the future and work to achieve it. Needless to mention the basic dual education in Arabic and French, encompassing writing, reading, and mathematics.

1.2 A Dog in Space (Breathless-Jerry Lee Lewis)

Without recovering from my first disappointment, I faced a second one shortly after. During a rainy day in November 1957, I heard that a dog called *Laika* was sent by the Russians (again) to space in another *Sputnik* satellite. I was excited by that piece of news and ran to ask my maternal grandfather for more details.

He was not very helpful, but he called me the day after to tell me that it seemed that the poor dog had died after reaching outer space. He said that this was the information all the accessible radio channels were reporting in their newscasts. His *Philips* radio set was able to pick up the Tunisian channel, the BBC, and some French channels like RMC.

Thanks to the previous explanation about the sky and outer space given by my paternal grandfather, I had some basic understanding of the tragic event, but without comprehending the whole story. My understanding was that in outer space, there is no air and Laika suffocated. I did not try to dig further, but, at the same time, I was sad and frustrated.

I was sad because the death of Laika reminds me of the death of my own beloved dog, a beautiful German shepherd called Fox. I raised this beautiful puppy myself since he was a few days old. Fox was a present from Dr Hoffmann, who was my ophthalmologist. I had a very serious allergy to pollen, which used to cause me burning eyes and near blindness in the midst of the crisis that occurred every year between the end of April and early June.

In the spring of 1957, my suffering was horrible, and the news about my mother's health might have further contributed to the aggravation of the situation. As usual, my father took me to consult with Dr Hoffman and get my cortisone injection. Before leaving the doctor's office, he put his hands on my shoulders and asked me if I would like to have a dog. I answered spontaneously, *yes*, because it was my dream to have a puppy.

He left us for a few minutes and came back with a small wicker basket covered with a kind of towel, on which a little dog was sitting. Dr Hoffman explained to me that it is a German shepherd, aged ten days, coming from the litter of his own female dog. He gave me a few pieces of advice on how to feed, clean, and educate the little dog. I had to give him a name. I chose *Fox*.

Shortly after, in early June, we came to our village, and of course, Fox was part of the expedition.

In September, the authorities decided to launch a nationwide campaign against rabies, and security forces started culling all errant dogs. One day, Fox was coming back alone from my grandfather's farm, and he was by mistake considered an errant dog. Consequently, he was shot dead and thrown into a bin. I was devastated by Fox's death and swore to never own a dog again.

My pain for Fox's death was still vivid when I heard of the death of Laika. This contributed to creating a sort of fear of everything linked to the sky or to space. In one word—up there, it is *dangerous*.

1.3 Death from the Sky (The End-The Doors)

On 8 February 1958, in the middle of the afternoon, we were in our apartment in *Sfax* when my father, who was an officer at the Tunisian National Guards (equivalent to the French Gendarmerie), came back suddenly from his office at the National Guards barracks.

He rushed to the master bedroom and collected his battle dress, his helmet, and different items he usually takes when he leaves for special assignments (potential combat mission, riot, arrest of dangerous criminals). My younger brother and I were afraid because we knew there was danger, and with the condition of my mother—who, meanwhile, had been told that her life expectation was relatively short we could face the risk of becoming orphans from both sides.

After being harassed by our questions and after swallowing some food, he answered us briefly:

"The French Air Force has attacked the town of Sakiet Sidi Youssef, near the Algerian border, killing dozens of people and injuring hundreds. All the casualties are civilians, with many children, because a school was targeted. The victims were both Algerian refugees and Tunisian citizens. As a consequence, and to avoid that the French army, which was still present in many locations scattered across the Tunisian territory, used this sensitive moment to redeploy its presence in the country and renege on the agreement signed in 1956—which led to the independence of Tunisia in March 1956—the National Guards have been tasked to control the French troops around their actual locations and to stop them from moving outside the allowed perimeters."

I believe that I did not understand the full explanation at that time, but one thing struck my mind:

"Again, death is coming from the sky."

This event of Sakiet Sidi Youssef became a national drama, and I was exposed to its narrative many times over the next few years, especially in school, where a memorial day was celebrated on every 8 February. On these multiple occasions, I understood that the attack was triggered by a desire for retaliation from France against the multiple actions carried out by Algerian freedom fighters in Algeria from Tunisian territory.

The consequences of this attack were multiple and seriously affected political life, mainly in France. It led to the deadly battle of Remada in the deep south of Tunisia on 22 May, between the French Army and the Tunisian army, backed by volunteers and the National Guard. This battle was caused by the *desperado's* attitude of the French generals, who initiated a putsch in Algiers on 13 May to show their discontent with the 4th Republic politicians.

I think it is important to mention that following the Sakiet Sidi Youssef attack, Bourguiba, the Tunisian president, wrote a letter to Eisenhower, the US president, complaining about the French actions and asking him for help. Eisenhower, who strongly supported the independence of Tunisia, did not hesitate to send a harsh letter to the French prime minister.

This letter generated a big row in France because it internationalised the Algerian war, against the will of the French authorities. This led to the resignation of the government of the French prime minister, Gaillard, on 15 April, then to the so-called Algiers putsch on 13 May, and finally to the arrival of General De Gaulle to power in the following two weeks.

During these troubled times, the French generals decided to ignore signed agreements and move more freely within Tunisian territory. This led to the previously mentioned battle of Ramada, in which my father was involved, generating a huge stress within our family.

Again, a nasty encounter with aircraft—already influenced by the stories of my family about WWII. For some time, aircraft were, in my mind, a tool of death and destruction. This impression was further magnified two years later by my encounter with comic books, one of which, called *Battler Britton*, was dedicated to aviators of WWII.

All the episodes were full of fights, sirens, bombs, explosions, destructions, deaths, and injuries. To be fair, I became a regular reader of these sorts of comics, but I was always troubled by the level of violence contained in the books. The majority were published in the late '50s and early '60s and were a direct description of what happened in WWII.

1.4 The Monster (Steppenwolf)

I continued to be haunted by the images of death and destruction provoked by aircraft until the mid-'70s—from the battle of Bizerte (July 1961) to the Six-Day War (June 1967) and the Yom Kippur War (October 1973), both between the Arabs and Israel—not to forget the Vietnam War, culminating with the image taken on 8 June 1972 of the little girl, Kim Phuc, running naked while the napalm was still burning her back.

At a certain point, I really started hating any military aircraft and stopped including more of this type in my database, which I had started filling with all the aircraft types I came to discover through direct viewing, movies, magazines, and comic books.

Furthermore, the intensive bombing carried out by the US in Vietnam, with the frequent utilisation of napalm and other defoliants, added to all that my father told me about his experience in Indochina between 1947 and 1949, all this mixed with my growing anti-imperialist beliefs, made me become an anti-Vietnam War activist from 1968, when I participated to the famous demonstration of March 1968, which triggered a wave of protests against the Tunisian regime and the US.

Later, in France, I participated in several activities organised by *Le Front Solidarité Indochine (FSI)*, which means Indochina Solidarity Front. This participation had unpleasant, but also pleasant, repercussions on my future life.

Although I had some sympathy for what is called the left of the political spectrum, I did not support this organisation, which had been created and was dominated by leftists. I participated in some events organised by FSI simply because I had a strong gut feeling that this war was not justified.

Above all, my father's story of his military involvement in Vietnam had a huge influence on my perception of the consecutive conflicts in that country (with the Japanese, the French, and the Americans). I was and I am still having inclinations toward the moderate left, and I never accepted Communism and all its offspring.

The Russian tanks killing a Hungarian in Budapest in 1956 (I came to know about this event through my father) and the immolation of Jan Palach in Wenceslas Square, in Prague, in 1968 (I followed the events personally) sickened me and played a repulsive effect vis-à-vis the Soviet Union and all the communist countries.

It is my firm conviction that imperialism and colonialism are not acceptable, and I will always stand by the side of the oppressed, whoever the oppressor is.

I was immersed in the late '60s and early '70s in the new wave of protest songs in the Arab world, Europe, and the US. It seemed that everywhere, there was a desire to express frustration and revolt through music and lyrics.

At first sight, everything seemed positive. The world was experiencing strong economic growth following WWII. Households in Europe and the US were experiencing a big improvement in their daily lives, and third-world countries were enjoying some relative improvements. However, it seemed that the young generation was feeling a strong need for more freedom, for big changes in the society and, in certain cases, for more democracy.

The Vietnam War, combined with the civil rights struggle was a good mixture to radicalise the protest movement. Artists like Bob Dylan, Joan Baez, Steppenwolf, Jefferson Airplane, and many others in the US; Jean Ferrat, Paco Ibañez, Léo Ferré, Guy Béart, Georges Moustaki, and many others in Europe; and finally, the numerous disciples of Sayed Darwish and Marcel Khalifa in the Arab world all stood up to revitalise the protests with powerful songs that left their mark on their respective countries.

Until today, people in Tunisia continue to sing revolutionary songs from 1968, 1969, and 1970. Some American and French songs are still popular today—*Born to be Wild* by Steppenwolf or *Le Poète a Dit la Vérité* by Guy Béart seem as if they were written today.

This music contributed to shaping my personality and my convictions.

Luckily, later, I had the opportunity to experience more peaceful and happier events related to aviation which allowed me to have a more balanced view about this important industry called Aerospace.

1.5 Aircraft Noise (Beautiful Noise-Neil Diamond)

In September 1960, we moved to Tunis to enable my mother to follow special treatment. Initially, her life expectation was said to be around six months, as stated by one doctor in early 1957, but her survival for more than three years encouraged her new doctor, a young professor in oncology, to suggest a new therapy protocol that could help reduce the size of the tumour without using surgery, which had become very risky.

During the last three years, I became very close to my mother because I was the one who had to describe to her everything she could not see. I had to develop skills of visual observation as well as of auditory interception. I had to describe all the details briefly and precisely. For this purpose, we developed a sort of secret code known only by both of us.

Since she also became very sensitive to all sorts of noises, I had to understand what she was describing to tell her what the origin was. The role of *visual translator* for my mother helped me a lot to develop my ability to observe things with more attention and with more focus on the most relevant things.

Our new home, which was located at El Mensah, was about some two miles from the runway of El Aouina Airport (now Tunis Carthage). This exposed us to the noise of all aircraft moving in and out of the airport.

There were not too many movements, but depending on the direction of the wind and my presence or not in open air, I could clearly hear the noise of the engines. I had a gross idea of what kind of aircraft it was, but I could not tell the difference between one type and another.

In 1960, the majority of aircraft were powered by propellers, and the dominant types were the DC3 and the DC4. I did not know anything about airlines, airports, or any other aspect of aviation. The only thing I knew was that there was a pilot who flew an aircraft. It was *Battler Britton* who gave me the basic knowledge of aircraft and introduced me to the world of aviation. Again, it is the military side which got me acquainted with aviation.

My knowledge of aviation made a quantum jump in March 1961, during the spring holidays. We used to go with our bicycles to the *stade vélodrome* of El Menzah, where we could race using the cycling ring of the stadium. The stadium was less and less used for football competitions. There was a security team, but they were flexible with us if we did not step on the football ground.

The cycling races were allowed, and the employees of the stadium were even watching us and sometimes helping us inflate the tyres and fix some mechanical problems. One day, in the middle of a heated race, an aircraft took off and generated a lot of noise. Caught by surprise or just by coincidence, one of our friends lost control and fell, causing himself a serious injury in his left arm.

We stopped the race, and one of the employees of the stadium, who seemed to have some medical knowledge, tried to do what was needed, but he did not seem to be happy with the outcome. Therefore, we went to the local pharmacy, and the father of our friend was called. He took his son to a close-by clinic where he received the necessary treatment.

We waited for their return, and later that day, all of us went to the house of our friend to inquire about his condition. His father cross-checked with us the circumstances of the accident and was surprised when we told him that the noise of the aircraft was the cause.

Since he was one of the handful of Tunisian engineers who were working at the airport, he laughed and told us not to be afraid of aircraft noises. He promised to take us all close to the airport the next Sunday to show us aircraft taking off and landing and to hear their noise from a short distance.

This happened as promised; and it was the first time in my life that I could see a DC3 and a DC4. Both aircraft were operated by Tunis Air, the airline of Tunisia. I could also discover the existence of Tunis Air and Air France. Our friend's father was very patient in explaining the difference between the DC3 and the DC4 (2 and 4 engines) and their respective seating capacities.

He explained to us that drinks and food were served on board. Since that day, I understood that aircraft could be a tool of peace, transporting passengers and cargo, linking cities, connecting people, and generating business.

During this visit to the vicinity of the airport, we also discovered a French aircraft called the *Nord 262*. It triggered a lot of curiosity among our group because *Nord* means north in French, and a lot of us believed that this aircraft was always flying to the north. We asked our friend's father, who burst into laughter and explained that it was just a name and the aircraft was able to fly in all directions.

I started having a more balanced view of the role of aircraft in our world. I also became more familiar with their noises and progressively started distinguishing between the DC3 and the DC4.

The situation was comfortable up to the arrival of new players with a new kind of noise: the Caravelle and the B707 started operating in Tunis, as well as some other Russian aircraft. It was easy to distinguish between the jet aircraft and the propeller aircraft, but segregating the jet aircraft was a slightly more difficult exercise.

Except for the two years spent in Sousse, far from any airport, I spent the rest of my life, up to the age of 18, close to El Aouina Airport, within a range of two miles from the runway. By 1969, I became very familiar with most of the aircraft types operating at the airport and could distinguish them by noise and shape. This was facilitated by the situation of our new house in Ariana, which was located close to the threshold of runway 11/29.

I could see aircraft very closely and recognise their main features. I saw the arrival of newcomers like the Lufthansa B727, the Comet-4B of British European Airways, which took over the London-Tunis route from British Eagle in 1968, the BAC 111 of BUA (later British Caledonian), and many others.

1.6 A Man in Space (Rocketman – Elton John)

A few days before my 9th birthday, I was playing football with some of my friends next to our school in El Menzah. It was close to 4 pm, after we left the school. The ball was shot wrongly towards the school and finished its course in front of the main door.

When trying to recover it without attracting the attention of the headmaster or some of the teachers who were still in the school, we failed and had to listen to a sermon from one of the teachers, who was kind enough to give us the ball and to inform us (probably as a matter of lesson to contemplate):

"Hey boys, do you know that a man flew to space? He is Russian, his name is Gagarin, and he did a full circle around the Earth in a vessel called *Vostok1*."

I don't believe I got the full message at that time. It was 12 April 1961. The event was huge, and I had the opportunity to learn more and more about it, including from our teacher, who did his best to give us an explanation with drawings on the blackboard and a simulation around the globe positioned on his desk.

The funniest explanation was given by Mr Zitoun, the owner of the local bar.

He did not stop repeating the same story:

"For a man to accept flying in a very hostile environment and in a canned box of sardines, he must be full of alcohol. It is easy for the Russians to drink vodka, and Gagarin must have swallowed at least two litres of vodka before flying to space."

For me, it was an extraordinary day, and I was impressed by the ability of the Russians to send everything into space—first an empty shell (*Sputnik 1*), second a similar shell with a dog, and then, third, a bigger shell with a human who could survive the experiment. My pride was even bigger because my family has some Russian origins and, more precisely, from Crimea.

According to family records, we come from the city of Balaclava, which our ancestors left for Istanbul during the period when the Ottoman Empire was controlling Crimea, around the middle of the 17th century.

The last link of our family with Crimea was the participation of my great-grandfather and his brother in the Crimean War from 1854 to 1856. My inclination toward Russia had nothing to do with communism, which had no chance to penetrate our family, which was economically liberal and socially conservative.

Honestly speaking, I was totally ignoring all the American experiments. The media was reporting the achievements of the Americans, but I did not have much interest. I had no special resentment or any bad feelings against the US, but for me, they were just challengers trying to imitate the achievements of the USSR.

I strongly believe that Gagarin was the trigger for my strong interest in space and astrophysics. This passion developed according to several axes in the following years: a passion for the universe and its structure, a passion for telecommunications through satellites and moving vehicles, and a passion for aviation.

All my life was consequently dedicated to finding a way to cope with these three passions at the same time. It took me a few years to understand the difficulty of pursuing all three tracks.

The Gagarin episode also triggered a sort of childish metaphysical turmoil in my mind: Did he see God? Did he see heaven and the angels? These questions caused embarrassment for my father, my mother, and other members of my family, and very quickly, I stopped asking them.

The other outcome of the Gagarin flight was that I started apprehending the difference between the sky and outer space. The sky is the place for aircraft, and outer space is the place for satellites. All the other aspects, whether physical, geometrical, chemical, or astronomical, were unknown to me.

The challenge was how to follow, simultaneously, what was happening in the sky and in outer space. One easy and enjoyable solution was to go and buy as many comics as I could afford with my modest weekly cash allowance. I started buying *Battler Britton, Superman* and the like. I also became interested in more complex books like *Le Petit Prince.*

Somehow, I was encouraged to read, and this became a passion. Some books were difficult for me to understand at certain ages, and I was obliged to revisit them once or twice after some time. By the age of 14, I had a solid knowledge of aviation and a modest but clear understanding of space and the universe.

2. A Teenager Passion
(Teenage Rampage—Sweet)

Having acquired a relative awareness of the sky and the space above our heads, I started developing a certain curiosity for every flying object. As soon as a noise was audible, high in the sky, I used to promptly try to locate and identify it.

In the beginning, the game was easy: just distinguishing between natural noises and artificial noises, particularly those emitted by aircraft. But things became more complicated when I started trying to identify the aircraft by their respective noises.

But destiny was to help me in my quest, as I had the opportunity to live close to an airport and hence get acquainted with aircraft, their shapes, and their noises.

2.1 Living Next to an Airport (Airport—The Motors)

After the death of my mother, in April 1963, in the city of Sousse, where we were based for the past two years, my father took the decision to move to the National Guard's headquarters in Tunis, where he was offered the position of logistics officer. Consequently, in September 1963, we came back to El Menzah, to the same apartment and the same school.

Of course, we came back without my mother, nor my grandmother, who had been visiting us quite often to help my mother. This time, my aunt Khadouja volunteered to come with us and help us during this difficult transition period. My aunt was the closest person to my father. She was his beloved sister, and each one could count on the other's support.

Later, we moved to bigger apartments—first when my father married Badra, the lady who became our second mother and gave us all her love, and second when we became teenagers and needed more space. But we stayed within the two-mile radius of the airport, with an even better location when we moved to Ariana. We were one mile from the threshold of one runway (11/29).

During my two years in Sousse, I had a headmaster, Mr Belaid, who gave me one of the most important lessons of my entire life. Due to my relatively good level of French, it happened a few times that I interrupted the teacher to correct the spelling or the pronunciation of a word. On one occasion, the teacher came to my table and gave me a slap on the face. Without any hesitation, I left the classroom and went straight to the office of the headmaster.

I told him my story. He looked at me, came closer, took my left ear between his fingers, and said to me gently and slowly:

"My son, when you say that you are always right, you become the fool of the village. Never forget this."

He took me by my hand, brought me back to the classroom, said something to the teacher, and left without any further word.

As soon as I came back home, I told the full story to my father and asked him for an explanation. To me, I had not done anything wrong, and I had no pretension to brag that I was good at French. I simply felt obliged to show the teacher that the spelling was wrong.

My father explained to me that the teacher was upset because he was supposed to teach us how to write, and he found himself taught by one of his students. This was not an easy situation for him, and he had felt obliged to show who the boss was. This is the reaction of the people who don't like to lose their leadership and/or acknowledge their mistakes.

"This was not fair to you, but this happens in life. This does not mean that you must shut up when you hear or see wrong things. But unfortunately, if you react quite often and if you are right, people will not like it, and they will try to corner you as being a special person—like the fool of the village. The right recipe is to be humble, not to show off, and to react strongly only when the cause is worth the action."

I was shocked, and it took me time to absorb all these subtle explanations, but this became a motto of my life and helped me a lot during my subsequent professional life.

Thanks to my growing passion for aviation and the opportune proximity to the airport, I became a passionate aircraft spotter. One of my favourite games was to bet with my friends on the type of aircraft landing or taking off. Over the years, I was able to enrich my database about the types of aircraft, their capacities, their range, their engines, and destinations from Tunis.

Strange enough, this passion did not create in me any desire to become a pilot or even work in aviation. On the contrary, I started developing a real love for radio communication and enrolled in *Eurelec*, an education programme via correspondence.

Thanks to this programme, I could manufacture, by myself, a certain number of devices, and I learnt a lot about telecommunications—starting from the technology of lamps and moving up to the technology of transistors and integrated circuits.

For six years, I was studying at Collège Sadiki, a reputable high school that has seen the graduation of hundreds of famous Tunisian politicians, civil servants (including my grandfather), and businessmen since its creation in 1875. During my years in this venerable institution, I had the chance, as in the several primary schools I attended, to enjoy a high-quality and modern education which was comparable to those provided in Europe.

We had teachers of high calibre who had an important influence on the students. Some of them deserve a special mention, like Mr Haddaoui, a teacher of history who, thanks to his narrator approach, made me love this subject and integrate it into my daily life; and Mr Tron, a teacher of mathematics, who was a genius with a strong charisma that enabled him to make complicated problems look simple.

By 1969, I had confirmation of my three strong passions: telecommunication, aerospace, and the structure of the universe. I could add a fourth one: Rock & Roll and music in general, which became an important part of my daily life.

I was balancing between Tunisian songs, with one main focus on two artists (one lady: Saliha, and one gentleman: Hedi Jouini), folk songs, Egyptian and Middle Eastern songs (Abdul Halim, Fairouz, Abdul Wahab, etc.), and European and US songs (everything, especially if it was Rock & Roll).

The first one was my favourite for my career, the second one was my preferred field for technical curiosity, and the third one was my field of intellectual curiosity—particularly everything linked to the creation of the universe and the galaxies. Music was a passion and somehow a physical need to cope with all the pressure around me.

I had the feeling that aviation was a sort of entry door to space and everything beyond. But somehow, I discounted aviation compared to telecommunication. This feeling was further strengthened when I learnt that it is thanks to radio waves (i.e., telecommunication) that we could detect objects in the universe. There is no debate: the future is for telecommunication. I kept my interest in aviation, but it was much more a hobby than a career aspiration.

By then, I had a solid knowledge of aircraft, and my database became huge. I had to dedicate a file for each country: US, France, UK, Russia, Germany, Canada, Japan, Poland, Sweden, Czechoslovakia, and the youngest comer, Brazil, with one subfile for military aircraft and a subfile for civilian ones. My headache was my inability to know if I had listed all the existing aircraft or not.

Each week, I came across an aircraft type which was not listed. It is true that it was a period of continuous discoveries, and the last seven decades were full of creation and innovation in aviation. I was simply astonished by the huge number of different types of aircraft built during WWI and later during WWII. I had the impression, maybe wrongly, that between the two wars, there were many more types of aircraft than types of cars.

It was a real frustration not to be able to know all the types. Some of them were built as prototypes only and did not last long. Finally, I decided to include only the aircraft types which were still flying by 1969. Luckily, some of my favourite ones were still flying, like the Breguet Deux-Ponts and the Convair880.

The Breguet was impressive with its double deck, and the Convair with its high speed. Unfortunately, none of them was a commercial success, but we must salute the engineering performances. I, humbly, could say that my future passion for the A380 could be rooted in my admiration for the Breguet Deux-Ponts.

Two British aircraft caught my attention, and I dedicated some time chasing all sorts of information related to them. The first and the oldest introduced multiple innovations (size, wingspan, automatic flight control, high-pressure hydraulic system, utilisation of electric motors in many automatic control systems), at an early stage of modern aviation. It was the Bristol type 167 Brabazon.

The second and the relatively more recent at that time was the Comet, which was the first jet aircraft dedicated to the transport of passengers with, again, a multitude of innovations.

I was impressed by the capacity of the British to innovate at such a pace, while the common belief was that the Americans were the most advanced in the field of aviation. Indeed, they were by the late '60s, thanks to the multitude of aircraft they had developed and built during and from WWII onwards.

2.2 The Beatles and the Pan Am 707 (Hold Your Hand-The Beatles)

In 1964, we had access, in Tunisia, to two TV channels: the national one and RAI 1, the Italian first channel. At 8 pm, on Friday, February 7, we selected, as usual, the Italian channel for a quick review of the main international news broadcasted by *Il Telegiornale de la Sera*.

This was for the adults, but for the children, the real interest was to watch the daily favourite short show called *Carosello*. That day, we were all surprised by the fact that the opening of the newscast was dedicated to showing a Pan Am aircraft leaving London airport, with the speaker saying that the plane was transporting the Beatles to the US.

I knew who the Beatles were and even knew a few titles of their most famous songs, which were frequently broadcast on the airwaves of the international Tunisian radio channel (Love me do, she loves you, I want to hold your hand).

The next day, being a Saturday, we finished school at midday and quickly ran to come back home, first for lunch, but mainly to be on time to watch a special programme for children, broadcast by RAI 1 after the 1 pm news. Again, the newscast dedicated a large part of its content to the Beatles, but this time, showing the arrival in New York.

The images of the crowd, the airport, and the plane were impressive. It was the first time that I heard the name *Pan Am*. I had already heard of the B707, but I could only see it on TV. The images shown on the Italian news were very special because they were associating two big names: Pan Am and the Beatles. For me, there was a third element: the Boeing 707.

At that time, I could not grasp all the implications of this special event, but I was fully aware that the Beatles were there to conquer the US public. I could not imagine what would happen later, and it took me several years before understanding the power of images: Pan Am, the Beatles, as well as, although to a lesser extent, Boeing, had fantastic advertising coverage for free. All the important TV channels of the world broadcasted this arrival in New York.

Analysed through my today's lenses, this was a masterpiece of PR exercise. Boeing was paid for the aircraft; Pan Am was paid for the seats used by the

Beatles and their accompanying staff, and the Beatles were going to generate a lot of revenues from their US tour. Pan Am was a simple service provider, thanks to the aircraft, which was a simple tool, but by the miracle of a combination of different elements, the airline and the aircraft became sort of heroes.

The result is that this aircraft is known until this day (serial number 17683 and registration N704PA, named *Jet Clipper Defiance*). In any case, for a teenager like me, who had a passion for aircraft and was progressively becoming a fan of Rock & Roll music, it was a perfect combination of my two passions.

I did my best to get a photo of this event by buying, over the next two months, any newspaper or magazine covering the event. I was lucky because a French magazine called *Salut les Copains* wrote some articles about the Beatles' tour in the US, in its March and April 1964 editions, with some photos.

This was not the only time I associated aviation and/or space with Rock & Roll. This was a sort of intellectual game I developed, by trying to find the possible link between the name of a group of musicians or a song and aviation or space.

There are many examples: Jefferson Airplane, Jefferson Starship (it took me time to understand the real origin of the two names, which apparently had nothing to do with flying machines), several Pink Floyd songs with titles related to space and astronomy (*Interstellar Overdrive, Set the Controls for the Heart of the Sun, Astronomy Domine, The Dark Side of the Moon*), for which I am still looking for a clear and convincing explanation, despite becoming a loyal fan of the group and reading a lot about its work.

2.3 First Close Sight of an Aircraft (In the Army Now-Status Quo)

From October 1968, I had to go to the military camp of El Aouina twice a month for military training, as part of a new programme introduced by the Tunisian Ministry of Defence, called *Youth Military Preparation.*

This programme was supposed to enable all young Tunisians reaching the age of 16 to fulfil their military obligation, in the frame of the national draft, through training sessions of half a day every fortnight during two consecutive school years, and through blocked periods of respectively 10 and 15 days during two consecutive summer holidays.

This was an enjoyable part of our school curriculum. First, we had the fantastic opportunity to have serious training on all military matters and on how to handle different types of weapons (pistols, rifles, machine guns). It also enabled us to be together, as students of the same class, outside the school, for several hours.

We could do many things we could not do in school: we could sing during the transportation by army trucks from our school to the camp, we could practice a lot of uncommon sports like obstacle course, we could practice shooting, and above all, learn some discipline which solidified the group and made us get closer to each other.

The trip to El Aouina, every two weeks, was a pleasant moment for me, especially since I had a golden opportunity to see the civil aircraft from a very short distance, as well as some military aircraft which were parked a few metres from our training field. The Tunisian Air Force was modestly equipped, and there were only a few small transport aircraft at El Aouina.

A big part of the Air Force assets was at the main air base in Bizerte. Even if they were small, the aircraft available at our camp enabled me to have a close look at the cockpit and to walk around the aircraft. The two or three aircraft we used to see at the camp were all powered by propeller engines.

I had the opportunity to have a close look at these aircraft thanks to the complicity of a captain who had been trained by my father in the late '40s when both were in the French army. He oversaw the security of the camp, and he got me special permission to stay after the end of the training session and took me for a tour until my father came to pick me up.

In a way, I had some privileges being the son of a veteran and a still active officer in the Tunisian National Guards.

I was somehow disappointed by what I saw. The aircraft were old and cramped, and their technology seemed very basic. It was the case of most of the aircraft there at that time, but I did not have a correct perception of the reality.

One other advantage of going every fortnight to El Aouina was the possibility to watch the Tunis Air aircraft moving in and out of the maintenance hangar, which was just located next to the military barracks and the air base. On several occasions, I saw a Caravelle towed out of the hangar. I had no clue what was happening in the hangar, but I could see the movements around it.

Following this close encounter with aircraft, I asked one of my closest friends, who was aspiring to become a pilot, to lend me a book he bought to get more acquainted with flying. I cannot say that the book impressed me. I simply was not interested in becoming a pilot or working in aviation. There was a big discrepancy between my passion for identifying aircraft and describing their specification and performances and a real desire to work in the field of aviation.

2.4 Walking on the Moon (The Police)

The summer of 1969 was one of the most exciting periods of my life. First, it was one of the busiest and richest in terms of lessons learnt, and second, it saw the achievement of my first set of radio receivers and different electronic devices built thanks to the correspondence courses of Eurelec.

Initially, my plan was to spend the first two weeks of July in my grandfather's house in Sousse and enjoy the nice beach of Boujaffar and the fantastic summer ambience of Sousse, especially the Aoussou Festival (celebrating the God Neptune) with its carnival on 24 July.

Before going to Bizerte, in the north of Tunisia, in the last week of July to accomplish my military duty for a few days and finish the training programme of the first phase of the military service called *Préparation Militaire Elémentaire (P.M.E.)* (Elementary Military Preparation), I planned to spend all of August in Sousse, while freeing some windows to attend a few concerts planned in Carthage and Tabarka during summer music festivals.

In the early days of July, I had just started enjoying my summer holidays when my father, who was joining us only during the weekends, came with a load of packages from Eurelec, containing all the kits and the instructions to build several electronic devices. I had been waiting for this delivery for some weeks.

Even though I was upset because the packages arrived at the wrong time, I was, at the same time happy and excited to be able to build my full set of equipment and make them operational as soon as possible. I was trying to accommodate my holidays, my military duty, the Eurelec project, and the concerts when I received another bombshell: Appollo 11 was to be launched very soon, with the aim to put the first man on the moon.

The mission was supposed to start on 16 July 1969 and last for more than a week. I was so excited by this news that I decided to cut short my stay in Sousse and come back to Tunis before travelling to Bizerte. This would allow me to spend as much time as I wished in front of the TV to follow the Apollo adventure and, at the same time, start building my electronic devices. In addition, I could have some windows to attend one or two concerts.

As of 10 July, I was in our apartment in Tunis with my father. I transformed part of the kitchen into a small electronic laboratory with several measuring devices, a soldering iron, and different utensils to assemble, screw, adjust, and rectify the equipment I had to build.

For a few days, I was able to swiftly advance in my endeavour to build my complete electronics laboratory, but from the day of the launch of Apollo 11, the lack of sleep and the need to catch up every two to three hours with the most recent developments of the mission slowed down my work for the laboratory, and my passion for aerospace became dominant.

My father was understanding, and he helped me by filling the fridge with basic food and bringing me different types of sandwiches when he came back from work. Neither of us was a good cook, hence our huge consumption of sandwiches and salads.

While my father was at work, I was juggling between my Eurelec stuff, Apollo 11, listening to music, and scarcely reading the small booklet given to me to help with the preparation for the military training session planned for 28 July. All this bathed in the Rock & Roll music of that time, which was going to have its first big festival: the famous Woodstock. I was enjoying my preferred music, living my two passions, and preparing for an exciting episode of my life.

I was missing the beach pleasures, but it was worth the sacrifice. The Apollo 11 crew became sort of relatives; they were part of my family, and I was focusing my attention on all their acts, reactions, feelings, and comments. I was frustrated with not being able to hear their conversation in real-time. I would have loved to see my amateur setup of electronic devices being able to pick up the Apollo 11 radio communications.

My dream was to meet these three guys and have a discussion with them about everything they lived through during their mission. I had to wait up to June 2010 to meet Neil Armstrong and have dinner with him and other people in one of the most post-war iconic aviation temples: Tempelhof Airport in Berlin.

I also had two opportunities, in 1983 and 1984, to meet with another astronaut, Charles (Pete) Conrad, who was the third man to walk on the moon, and I had an exchange of presents with Eugene Cernan, the 11th man to walk on the moon. These encounters will be further detailed in the coming chapters.

21 July was a memorable day. The first human step on the moon and the completion of my radio receiver (with some limited bandwidth but working). That day, I was also exposed to the first conspiracy theory in my life. On the

occasion of the Gagarin flight, I was the one asking silly questions, like: did he see God? Did he see the angels, paradise, or hell? Some of my friends were on the same line.

This was pure childish understanding of the event and was, somehow, normal. This time, some of my friends, many adults, and people from different ages, social origins, and religions were refusing the fact that Armstrong and Aldrin had walked on the moon. For them, it was not real—NASA made a movie on the ground. It could not be true because it would have been a sacrilege, and God would have punished them.

One of the most vocal ones was one of our neighbours, Mr David, who was a practising Jew. For him, it was a sin to provoke God, and therefore, the American scientists and engineers, who are in their large majority strong believers in God, cannot dare to do so. It was a Hollywood movie. He was strongly supported by his Muslim neighbour and friend, Salem, who was a teacher of Arab literature in a secondary school.

In the days following the landing on the moon, a heated debate was taking place between those who believed in the reality of the events and those who did not. I was personally surprised and shocked by this sort of debate and could not imagine that science and technology could be suspected of manipulating the truth. For me, everything coming from science was an absolute truth, not negotiable and not debatable. It was an Axiom.

It took me many years to shift to a more balanced and understanding attitude.

Finally, during this summer of 1969, I could cover all my centres of interest and fulfil my military commitments, with a memorable stay at la Caserne de l'Horloge (the Clock Barracks), where my father had been stationed during the '30s and got his training to become a sergeant.

I could also attend two or three concerts, but the only thing I could not follow was Woodstock, because of the lack of coverage by both the Tunisian and the Italian TV channels. Only some newspapers and some magazines allowed me to have limited access to this big event. I waited for more than a year to be able to buy the records related to the event and a little bit more to watch the movie.

My interest in the moon continued with the consecutive Apollo missions and the Russian Lunokhod or Lunakhod adventure, which was also an exciting saga due to the challenge of remote control from the Earth of devices moving on the surface of the moon, using the technology of the late '60s. To be fair with the Russians, it was an outstanding achievement.

Unfortunately, I had to slow down on dedicating a lot of my free time to my aerospace and telecommunication passions because I started a new cycle of education, which was, and still is, known to be very difficult and very demanding, and where every second counts.

From September 1969 to June 1970, I had a very hard school year, dedicated to the preparation of the national examination of the Baccalauréat, which is the sesame to access the university. I was lucky to be at Sadiki. I had excellent teachers, small classes, and a motivating environment.

Although the country was poor, we were getting a good quality education, and I had time to dedicate to my hobbies: aviation, Eurelec, Rock & Roll, and to following my favourite football clubs—l'Etoile in Tunisia and Manchester United in England. I had very busy weeks, to the point that, just a few days before the famous and feared examination, I had to shuttle with my friend, Sahbi Mahjoub, to Sousse to watch a game between l'Etoile and l'Espérance, two of the four biggest and oldest football clubs in the country. The game ended with fights and long delays in transportation to return to Tunis.

This caused great anger among our fathers, who called us reckless and crazy by risking our future because of a football game. My passion for football grew further, and I became a multi-fan of five clubs: l'Etoile in Tunisia, Manchester United in England, Olympique de Marseille in France, Juventus in Italy, and Real Madrid in Spain. But l'Etoile remains number one and Manchester United number two.

They were my first loves and remain so. I cannot forget the pleasure I felt watching the glorious teams of the '60s—Chetali of l'Etoile remains an icon, and in Manchester United, George Best was an attraction on his own. Of course, I'll never forget the style of Nobby Stiles and the charisma of Bobby Charlton.

I admired Manchester United for their game, for the personality of the players, but also for the capacity of the club to recover and rebuild its strength after the crash in Munich (1958).

Bobby Charlton and Bill Foulkes, who were two survivors of the crash, succeeded in winning the European Cup 10 years later. This is what is called iron will and perseverance.

2.5 First Air Travel with a Caravelle (Give Me a Ticket for an Airplane — The Box Tops)

In September 1970, I had to leave my family, my friends, my house, and my country to travel to France, and more precisely to Toulouse, to pursue my studies. Having obtained my Baccalaureate in mathematics in June 1970, and having obtained good marks, I was selected to be part of the group sent to France to attend the preparatory classes for the Grandes Ecoles.

I was assigned to Lycée Pierre De Fermat (a famous French mathematician who is the author of the famous last theorem of Fermat, which needed 300 years to be properly demonstrated) in Toulouse.

To reach Toulouse, I needed to take a Tunis Air flight from Tunis to Marseille and then a train from Marseille to Toulouse. The route was operated, at that time, by a Caravelle, and for me, it was a great source of excitement and curiosity to fly in this mythic aircraft.

I moved at least twice from my window seat during the flight, to go and see how the toilets, the galleys, and the cabin crew seats were. Regrettably, I could not enter the cockpit, but I could have a look when leaving the aircraft after arriving in Marseille.

I was impressed by the low level of noise inside the cabin due to the position of the engines, which was a real revolution in aviation and became a sort of standard for many aircraft to come (DC-9, BAC 1-11, Fokker 70, Fokker 100, TU-134, and within the same philosophy, the B-727, the Trident, and the Tu-154). A few years later, I had the chance to have a very close look at the Caravelle and discovered all its beauties.

This first air travel gave me an initial first impression of what is called the air transport industry; and in the early '70s, it was much more a craft activity, in everything besides the manufacturing, maintenance, and flying of the aircraft. Ticket sales, passenger check-in, boarding, and all the non-technical handling were done manually and were like the practices of ship, train, and bus transportation.

I was somehow disappointed to see the checking agent scrolling a list of names printed on several pages to find my name, check in my luggage, and give me my boarding pass. I thought the process would have been much more automatic.

I thought maybe Tunis Air and Tunis Airport were not advanced enough, but my impression was confirmed during my return to Tunis for the Christmas holidays, where I saw that Air France (the handling agent for Tunis Air in

France) and Marseille Airport were working the same way. It took several years before seeing the first IT-based and automated systems being implemented.

These first impressions about the air transport industry played an important role in defining my vision of my future career.

3. Student's Dreams and Challenges (Ô Toulouse—Claude Nougaro)

The arrival on French soil, in Marseille, was one of the very confusing events in my life. I was repeating what my father did 30 years ago when he arrived in the same city before moving to the north to fight for France against the Germans. But in my case, the objective was more peaceful. I was an alien coming to a country which was occupying mine, seeking more knowledge and planning to return to serve my country.

How would the French interact with me? Am I going to be confronted with racism? Am I going to adapt to the living conditions and to the climate? All these questions were whirling through my brain. As much as I was reassured by what my father told me, the recent events in France and the feedback from Tunisian immigrants equally worried me.

My journey on the train between Marseille and Toulouse was full of curiosity when discovering the landscape. I had the impression to be in the same position as my father, sitting in the car driving him from Marseille to La Roche-sur-Yon, passing through the southwest of France.

We arrived at the train station Toulouse-Matabiau around 5 pm. It was a sunny and warm day, and the rosy colour of Toulouse invaded all our visual field. Toulouse really deserves its name, *La Ville Rose—the Pink City*.

3.1 A Supersonic Encounter—Concorde (Speed of Sound—Coldplay)

I was properly installed in my individual cell, in the huge dormitory of Fermat, and classes started at a brisk pace. I did not have time to discover the city or to know more about its history, but during the last Saturday of the month of September, I decided to take half a day off to buy some additional clothes and other amenities.

While I was strolling through the streets of Toulouse, I saw a crowd surrounding a couple, with the man shaking hands and signing autographs.

I thought they were surrounding a famous singer or actor (I thought about Claude Nougaro, a famous singer who was from Toulouse). But one by-stander told me that the man was André Turcat, the chief test pilot of the Concorde, the famous supersonic aircraft. It is true that Turcat was taller than Nougaro, and I should have left Nougaro aside.

I became familiar with the Concorde, with its regular noisy over-flights of our school. We were warned by the teachers and the administrative staff that we would regularly hear the Concorde. The flight tests had started on 2 March 1969, under the commandship of André Turcat, and it reached Mach 1 (speed of sound) on 1 October 1969, under the commandship of Jean Pinet.

A few years later, I became friend with Jean Pinet, and we contributed to making a dream come true (the two-man crew cockpit in a wide body). Meanwhile, I continued suffering from the loud noise of the Concorde, especially when it took off from Blagnac airport, some seven km northwest of the Toulouse centre.

On 4 November 1970, just after All Saints' Day, Concorde reached the Mach 2 speed, and I remember a lot of wine and beer were consumed during that day in Toulouse and its suburbs.

I understood, then, that Toulouse breathes aviation and lives thanks to it; and that all the noises and all the side effects of the test flights will not deter the city from becoming the capital of aerospace of France and of Europe. I had the opportunity and the chance, a few years later, to contribute to this adventure both from the airline's and the manufacturer's side.

Toulouse made me also discover rugby and, in particular, the *Stade Toulousain,* which is an institution in the city and its suburbs.

Together with aviation, Toulouse also breathes rugby. This led me to learn and play rugby as a hooker, the only position I could play due to my comparatively small size. My short experience of close to four years cost me the dislocation of my left shoulder during a game against another school later on in Paris and led to an urgent operation. I still feel the consequences of this operation today.

I will never forget the way I was rescued by my friend and loose-head prop of our team, Francois Quentin, who carried me on his back for hundreds of metres before finding a taxi to take me to the hospital. I had the opportunity to see Francois again many years later when he was COO of Thales (previously Thomson), and he did not fail to remind me of my accident.

My arrival in Toulouse coincided with the start of the fierce fights between the Jordanian army and the Palestinian Fedayeen of the Fatah, a faction of the PLO. This deadly war, which lasted until July 1971 and killed thousands of people, in majority Palestinian civilians, was the origin of the creation of the terrorist group called Black September.

Because I was wearing a Palestinian keffiyeh, which I had bought on the occasion of a solidarity rally with the Palestinian civil victims, the students of the second year, who were organising the yearly classic hazing, decided to nickname me Fatah or Black September. I carried these two names for a certain time.

When facing some accusations that I was supporting the terrorist group, I decided to alert the Dean of the school, who reacted positively to my alert and called the class leaders of all second-year classes, asking them to stop calling me by these two names.

My action was well-timed because, later on, the terrorist group carried out deadly operations, like the hijack of the Sabena aircraft at Ben Gourion Airport in Israel and the Israeli hostage-taking during the Olympic games of Munich. I have always shown support to the Palestinian cause, and I continue to defend their right to have their own state in the frame of the 1948 UN resolution, but I cannot accept the killing of innocent civilians, from any side.

Unfortunately, Israel also continues to kill civilians, but nobody really cares, as it is mostly considered by the West as self-defence. If the West is seriously keen to have peace in the region, it must impose the implementation of all the UN resolutions and the Oslo agreement. Continuing to condemn always the same side and label it as terrorist does not solve the problem.

We must remember that the French resistance and the European partisans were labelled as terrorists by the Nazi regime. The same was applied to the Algerian FLN by the French authorities and to the South African ANC by the Apartheid government.

Mandela was accused to be a terrorist. Finally, the other side ended up negotiating with the terrorists, and definitive peaceful agreements were reached.

The world must not let the extremists from any side run the show and refuse any compromise.

I spent three memorable years in Toulouse, during which I prepared for the entrance examinations to different renowned engineering schools in France. It was hard, challenging, and sometimes frustrating, but I accumulated a qualitative quantity of knowledge in mathematics physics, and chemistry, which was second to none in Europe and even worldwide.

I also improved considerably my French and my English, to the point that I obtained the highest mark of 18/20, in all of France, in the French literature exam for *le concours commun*, to enter a group of schools (Ecole des Mines, Ecole des Ponts et Chaussées, Ecole des Telecommunications: Sup Telecom, Ecole de l'Aéronautique et de l'Espace or Supaéro).

This allowed me to be ranked at the top of the list and be able to choose the school I wanted to attend. It was a little weird that French literature was the decisive element in defining my future, while it was supposed to be the scientific subjects. But this is the name of the game in this type of competition: the secondary subjects are the decisive elements. Anyway, thanks to Diderot (the subject of the exam), I became the master of my destiny by having the possibility to choose my preferred school.

Supaéro was an option, but without any hesitation, I decided to choose Sup Telecom. I believe that all the subjects related to electromagnetism, relativity, quantum mechanics, and the structure of matter had a stronger influence on me than my teenage passion for aviation, which suffered also from its military shortfalls.

In addition, and while I was preparing for the oral part of the exams, I heard the news of the crash of the Tupolev 144 on 3 June 1973, at Le Bourget Air Show. This was, also, a contributing factor to my choice.

Toulouse and Fermat were also the source of a multidisciplinary education covering arts, music, history, gastronomy, knowledge of religions, politics, etc. The main thing I learnt, in Fermat, was that whatever the quantity of knowledge we learn, we are still ignorant of a lot more: what I know is a drop of water in an ocean (very well expressed by the song of Jean Gabin 'Je sais,' which says *what I know is that we never know*).

I attended many concerts of famous groups like Pink Floyd during their concert at the old *la Halle aux Grains*, in November 1972. It was the first time in my life that I got very close to the smoke of hashish which had filled the arena.

For a non-smoker like me, who, during his childhood, was sickened by the troop cigarettes (sold to the military personnel) which were heavily consumed by my father and which he made me try at the age of 13, with the logical consequence of rejecting them, this hashish smoke was disgusting.

Everything was psychedelic with Pink Floyd: the crowd was flying into space. I discovered also new sounds and new styles like those of Kraftwerk, the German band, and Magma, the French band. I watched *Woodstock, Easy Rider, 2001: Space Odyssey*, and several other movies which became true references of modern cinema. I read more about the two World Wars, the Spanish Civil War, the Vietnam War, the revolutions in Latin America, and many other important events.

I had the opportunity to understand the real French society and the clear divide between the fascists and the rest of the society. Toulouse was a leftist city, where socialism and left radicalism were dominant, but there were some families still nostalgic for Pétain and the 3rd Reich.

I had to argue and, even, physically fight against some students who were fascist and racist, and particularly one of them, whose father had been a collaborator with the Germans and was jailed in 1944 by the Résistance. He told me that I did not deserve to study in France and that this was a great favour. I answered that while my father was fighting against the Nazis, his father was licking their bottom; therefore, my family served France more than his family.

He was much taller than me and knocked me down with a direct blow in my face. I tried to react, despite my bleeding nose, but other students stopped us, and we finished, both of us, in the office of the General Superintendent, who happened to be a former comrade in arms of my father in the army. He quickly understood the situation, took harsh decisions against the other student, and then took me to the infirmary.

While walking, he put his hand on my shoulder and said:

"I am really sorry, France has still some remaining vermin. I know the value of your father and you are most welcome in our country."

I also discovered more about French food and beverages. I visited a large part of southwest France and I tested many restaurants. I fell in love with the city of Cahors and its surroundings, like Douelle, where I became a regular customer of its famous restaurant *Auberge du vieux Douelle, Chez Malique*.

I became familiar with Concorde, which I saw from the limits of the airport several times, and got first sight of the new Airbus A300B2, which started appearing on the tarmac as from 28 September 1972, when it was presented to the world, jointly with Concorde No. 2. Actually, at the time, no one really took notice of the *Airbus*.

I used to have a friend living in Blagnac, and we could use one of the balconies of their house located on a farm, close to the airport, to watch the movements of aircraft around the airport. Both the Concorde and the A300B2 became familiar silhouettes to me, which helped me a lot, a few years later.

Another great discovery in Toulouse was the Catholic religion. In Tunisia, I was living all my childhood in an open society where, as a Muslim family, we were interacting on a daily basis with people from other religions, and in particular Jews and Christians. The Jews were Tunisian nationals, we spoke the same language, we were often living next to each other, and we have very similar traditions.

The Christians were, in their majority, French nationals, living in their own neighbourhood and speaking only French. There were also some Italian and Maltese immigrants who were living next to us and who, in their majority, also spoke our language.

We were familiar with the major religious events of the other communities, but there were two of them that were very popular: one Jewish, taking place in May, the pilgrimage to the Ghriba, the oldest synagogue built outside the Holy Land around 586 BC in the island of Jerba. The other is Catholic, taking place on the 15 August in the church of La Goulette and called the celebration of the Madonna of Trapani.

These two events attract, until today, all the communities and are the occasion for joyful celebrations. As a Muslim, I knew a lot about Moses, Jesus, and the Virgin Mary, and I had to believe in their existence and their Gospel because these were conditions to be a Muslim.

The only thing I did not know about Catholics, at that time, was that they could represent God, Jesus, Virgin Mary, and the angels through paintings, statues, and other forms, all of which are called icons. Since the Muslims and the Jews forbid any representation of God, the prophets, and the angels, it was a real shock to see these huge frescos in the Covent of Les Jacobins or the Basilica of St Sernin in Toulouse.

My curiosity pushed me to join the Catholic community of St Sernin and participate in a lot of their activities, under the leadership of Father Jugla, who became my big brother and with whom I had my most interesting debates about

God and religions. These discussions contributed to forge my beliefs and convinced me that the three main religions are similar, have the same roots, and were highly influenced by pure human political considerations.

The core aspect (i.e., the belief in a creator) is the same and it even transcends the content of the various religious texts. I started discovering the fingerprint of the creator in many scientific aspects, and my question to a man of faith like Father Jugla was why the additional complications with all the legends and stories around the prophets and their status. I never had a satisfactory answer either from him or from his Muslim counterpart.

Therefore, my only solution was to read as much as possible of the Koran, the Bible, and the New Testament. It was a very rewarding exercise that gave me much-needed peace of mind and contributed a lot to shaping my personality. I became very tolerant concerning religions, with a strong mistrust vis-à-vis the non-tolerant.

I travelled from Toulouse to several places and, in particular, London and Paris, to attend concerts or demonstrations against the Vietnam War.

Every summer, I used to invite five or six friends to come with me to Tunisia. We used to take two cars and travel across Italy and catch a ferry from there to Tunisia. Thanks to my French friends, I discovered all the south part of Tunisia, down to the Sahara, and visualised all the famous spots where the fiercest battles had taken place between the Allied forces and the Axis forces.

There is a complete chapter of World War II about this, which needs to be told to the world. Luckily, there is one movie which answers part of this request: it is *Patton*, a movie which describes, with a lot of realistic details, the war in Tunisia between November 1942 and May 1943.

I was particularly impressed by finding, in the middle of nowhere in the desert, a wooden panel hanging on a door of a sort of natural cave with the German inscription 'Ostkommandantur' and a charred tank next to it. This was a few kilometres from the city of Tataouine, which gave its name to the famous planet Tatooine, and which is the home of Luke Skywalker in *Star Wars*. So, in the same location, you have a connection between WWII and Star Wars.

On 3 June 1973, I was studying some lessons and rehearsing with some friends how to deal with some tricky mathematical questions. It was late afternoon, and we decided to stop working and listen to some music. We were tuning our radio set when we captured the breaking news that the TU144, the Russian competitor of the Concorde, crashed at the Le Bourget Airshow.

Some journalists were nicknaming this aircraft *Concordsky*, insinuating that the aircraft was a simple copy of the Anglo-French Concorde. It was true that a lot of spying activity from the Soviet Union had taken place in France and the UK during the development of the Concorde, but it must be accepted that the Tupolev had some innovative and interesting features which are unique to this aircraft and made it different from the Concorde.

In this respect, I have noticed that the Europeans, and particularly the French, had a constant tendency to accuse the Japanese of copying their cars and the Russians their aircraft.

I personally did not believe in these accusations. First of all, the Russians have given the world eminent scientists like Nikolai Zhukovsky, who can be considered as the father of modern aerodynamics and had founded the first laboratory dedicated to aerodynamics in Europe, the TsAGI. He elaborated the famous theorem of Kutta-Zhukovsky, which contains his name and the name of his partner in the research, the German scientist Martin Wilhelm Kutta (this theorem shows how to compute the lift generated by a flow of a fluid around a profile).

Therefore, the Soviets were able to create and innovate. Sputnik and Gagarin had shown their capabilities in the aerospace domain. The Japanese had also the capability to innovate, thanks to their mastery of several modern technologies. I personally had the opportunity to be exposed to NEC leadership in microchips as early as 1976.

I was shocked by this accident, because I was a strong believer in supersonic passenger flights, and Concorde and Tupolev were the showcases for it. I was afraid that any accident would be a good excuse to stop the supersonic adventure.

Unfortunately, that is what happened ultimately, but I did not suspect that the oil price would be the real killer. As mentioned previously, this accident was also a contributor to my decision to choose telecommunications for the second phase of my university studies.

I left Toulouse in July 1973 to go to Paris and prepare for a permanent settlement there, as from October 1973. Within two weeks, I completed all that was needed and flew back to Tunis to spend some good time in my favourite place: Sousse.

3.2 Loss of a Close Friend in an Aircraft Accident (Stairway to Heaven — Led Zeppelin)

I came back to Paris around 15 September and took a room in the Residence of Tunisia, Boulevard Jourdan, not too far from the Telecommunication School.

The study programme was encompassing several scientific fields, ranging from probability to electromagnetism, to the structure of matter, to transmission, satellites, etc. In an avant-garde approach, the school dedicated a good chunk of time, for those who were interested, to quantum mechanics and solid-state physics, to tackle the growing role of transistors and processors.

From the start, I became fond of quantum mechanics and left aside aviation. It was a real intellectual pleasure to study this strange branch of physics, where nothing is sure and all is probability.

While I was in my first months in Paris, I received the shocking news of the death of one of my closest friends, Abdessattar Allal, who was a student pilot at the aviation institute of Tunisia called Aviation School of Borj El Amri. He had finished all the required basic training, and he was flying with one of his colleagues without any instructor. I cannot remember what the purpose of the flight was, but it was something beyond the mandatory training.

The weather was not very stable, and they faced a storm. They had difficulties to manoeuvre to get out of it, and they fatally crashed. I did not dare to ask more details about the circumstances of the accident. For me, the death of my close friend was so devastating that no explanation was worth being requested.

I started viewing aviation with some fatalism. This was the straw which broke the back of the camel. I decided I did not want not to hear any more about aviation, until further notice.

3.3 Paris, The City of Light (Paris s'éveille — Jacques Dutronc)

The life in Paris was much easier than in Toulouse. First, I had much more money. Second, the workload was lighter, and third, Paris offered multiple opportunities for entertainment and leisure. I was lucky to have part of my Toulouse gang with me and to reunite with my two best friends from Tunis, Sahbi Mahjoub and Hassen Khelil.

I moved outside the Tunisia Residence to an apartment located rue St Honoré, which I shared with my close friend Michel Forgue. This apartment was in the heart of the city and offered me the opportunity to reach any interesting spot of Paris within a walking distance.

In Paris, I logged dozens of rock concerts and movies. I tried to see all my favourite rock bands and singers. I also tried to improve my political education by attending conferences and political meetings of different French parties and organisations from both sides (left and right). I attended several meetings of the FSI.

After one of these meetings, in May 1974, at La Mutualité, a Tunisian opposition group asked the attendance to remain in the room. The purpose was to give a briefing about some prominent political prisoners in Tunisia, and especially about Gilbert Naccache. I was among many young Tunisians who were feeling the regime was too oppressive.

I decided to stay and listen. Some persons, who were part of a group combining members of the Tunisian political police and some activists of the ruling party in charge of monitoring the opposition and disturbing its activities, were present at the meeting and succeeded to identify me.

Nothing happened, until I received a phone call, on 28 June, from my friend Sahbi, who travelled to Tunis two days earlier, asking me not to travel to Tunis in the coming days because I could risk being arrested for opposition to the regime. I stayed hanging around in Paris for close to three weeks during which I could attend some rock festivals. Finally, Sahbi, called me to give me the green light to travel back home.

It is funny to mention that many years later, I got introduced to at least three of the members of this anti-opposition group and we developed cordial relations.

This episode comforted me in my assessment of the political life in Tunisia and further pushed me to stay away from the Neo Destour, the ruling party of President Bourguiba.

I never became an adherent of this party despite a lot of pressure and even once the issuance of a membership card in my name without my knowledge, but I never paid the contribution, and I tore the card up when I found it on my desk in my office at the Tunis Air maintenance centre. This attitude influenced the future development of my career. I was considered a free electron with no political backing.

With the signature of the peace accord between the US government and the three Vietnamese stakeholders on 27 January 1973, under the leadership of Henry Kissinger and Le Duc Tho, and the dwindling interest from all sides in the FSI, I stopped attending rallies organised by the FSI by the end of 1974. Nevertheless, I continued my activism against the oppression in Tunisia and I had to dodge with some risks.

3.4 Engineering Dreams (Sweet Dreams — Eurythmics)

From the first day, I walked into Fermat, I started dreaming that I would become an engineer, even if I was still hesitating between some fields of activity. The fact that I started my courses at Sup Telecom made my dream come true. Theoretically, and if I would not behave stupidly, I should get my diploma at the end of the three years of studies.

In particular, I owe a lot to my teacher of physics and chemistry, Mr Troiplis, in Mathematics Special (which is the second year at Fermat). He succeeded in giving me a clear understanding of the articulation and interaction between the different fields of physics. In particular, he opened wide the door for electromagnetism and telecommunication.

The only question was:

"What will be my field of activity?"

During the first two years in Sup Telecom, I had to do some internships. I found a good combination by splitting the allowed time between France and Tunisia. The direct result was reducing the length of my holidays, but it was very rewarding, in terms of lessons learnt.

In France, I had the chance to spend some time at two different locations of the CNET (National Centre for Telecommunication Studies), which was the research body of the French state covering all the aspects of telecommunication. During the two sessions I spent at the CNET, I discovered how developed countries prepare the future.

In Tunisia, I worked as a simple technician, once in an Ericsson switchboard centre (at that time the Swedish company was the world leader of cross-bar switches, and they were equipping many countries all over the world). Another time, I was learning how to troubleshoot transmission lines (using cables or using antennas and radio waves).

In the two sessions I had in Tunisia, I learnt how to deal with immediate day-to-day challenges. But in both countries, I learnt the meaning of teamwork and time pressure, and how important is to respect the midday break for lunch (even if it was a sandwich).

These lessons were of great help when I started working in the technical directorate of an airline. But they also convinced me that I would never work as an engineer in a telecommunication company, dealing with daily operation and maintenance. I had to go to the field of research where I could develop new concepts and new equipment.

3.5 Information Technology is the Future (Communication Breakdown — Led Zeppelin)

My connection with Eurelec gave me an opportunity to become familiar with electricity, electromagnetism, telecommunications, and components. The fact to learn while building something was an incredible efficient method of teaching. When I joined Sup Telecom, I found myself in a familiar environment and could rapidly increase my knowledge.

Very quickly, the link between telecommunications, computing, and solid-states physics became obvious, and I started hearing new expressions like field-effect transistors, micro-processors, RAM, ROM, wave guides, optical fibre, signal theory, information technology, etc.

I noticed that this world was governed by probability, and I still remember a funny name (Borel Tribe), which was one of the fundamental elements underlying signal treatment and other IT aspects, as well as probability of presence, which was one of the fundamental elements of quantum mechanics.

I discovered that I was navigating in a sea of uncertainty, but nevertheless, one could find his way, thanks to a clever utilisation of mathematical operators. It was possible to switch from a theoretical world to a semi-real world using some intelligent keys. Intellectually, this was fascinating, but it was also frightening.

These were my first steps in the world of IT, and I had to define the way I'd like to progress in it. I found the answer very quickly. It was interesting to study all the aspects of the world of IT (electronics, signal theory, transmission, networks architecture), but for me, I had to focus on the core of the system, and for me, it was quantum mechanics.

3.6 The World of Quantum Mechanics (Atom Heart Mother — Pink Floyd)

Without any hesitation, I decided to deal with quantum mechanics. Very quickly, I got attracted by this theory and I became convinced that it was going to put me on the track of semi-conductors and micro-processors, which were the rising stars, in 1973/1974. I literally dove into this field of studies.

The beauty of quantum mechanics is that it defies the common understanding and brings to light so many new concepts like the Heisenberg uncertainty principle, which, in a simplified manner, stipulates that we cannot know simultaneously the position and the speed of a particle; and the Pauli exclusion principle, which stipulates, for example, that electrons belonging to the same system cannot be in the same quantum status. The quantum entanglement is a phenomenon showing that two particles, which are part of a linked system (to simplify: they are members of the same group), will evolve and get quantum status depending on each other, whatever distance is separating them. It means there is a strong correlation between the physical characteristics of these distant particles.

With all these strange properties, I started digging into the quantum mechanics world with a mixture of excitation, admiration, and confusion. Very quickly, I started asking myself if there was a metaphysical aspect in this theory. It, somehow, contradicts, at least at first sight, the relativity theory of Einstein which was praised as the mother of all theories explaining the complex functioning of the universe.

I have no pretension of commenting the relativity theory, but I found it beautiful, especially when you have a good teacher who introduces you to the basics, starting from Electromagnetism and the Maxwell theory.

I must confess that I found it very hard to apprehend all the new aspects I was encountering, but thanks to the powerful mathematical tools used in quantum mechanics, you get a sort of operating manual that helps you play with all the concepts, without grasping the real physical world underlying the theory. It needs serious questioning of all acquired knowledge to be able to reach a modest understanding of this theory.

I really admire the founding fathers of this theory: Erwin Schrodinger, Werner Heisenberg, Wolfgang Pauli, Paul Dirac, Niels Bohr, and all the others, for their ability to conceptualise such complicated phenomena and come up with brilliant explanations.

It is important to give this kind of people the recognition they deserve for their contribution to the whole of humanity. I was very pleased and happy when I heard that Alain Aspect, John Clauser, and Anton Zeilinger got the Nobel Prize in physics for their work on quantum entanglement.

Like relativity, quantum mechanics has an impact on our daily lives, and it is difficult to imagine that some apparently simple equipment we use every day is an application of the properties of quantum mechanics.

I was proud and excited just to discover this fantastic theory, which, literally, impacted the way I apprehend life in its multiple dimensions. I was helped and supported by my teacher, Mr Ayrault, who was very close, available, and dedicated to his students. For me, quantum mechanics also became part of my quest to understand the big picture and try to have the beginning of an answer to the question of the creation.

My immersion in quantum mechanics did not make me completely forget aviation (which I continued to follow, through some magazines and during my frequent visits to Toulouse), and I could have some time to continue reading about the universe and other subjects linked to astronomy and astrophysics.

In the middle of these exciting times, where I was discovering new things every week, we were hit by the shocking news of the start of the Lebanese civil war in April 1975. We saw the devastating effect of this news on our Lebanese friends.

I was very close to a Lebanese Christian student, Jean Jacques Davidian, who was with me in the same school, and I was updated on a regular basis. All our group of friends were plunged into the details of this stupid war, which was the consequence of multiple political blunders by the Lebanese themselves and by many foreign countries. This war was part of my daily life for some time and created a special bond between myself and Lebanon.

My passion for quantum mechanics made me work hard and try, as much as possible, to get a scholarship in order to pursue my studies in one of the top US universities with a renowned department in quantum mechanics and solid-state physics because I needed to see concrete application of the theory.

My efforts were rewarding to the point that, in June 1975, I was offered the possibility to move to the US as of September 1975, precisely to Texas A&M, with a respectable scholarship, to prepare for a PhD in solid-state physics.

I got everything ready, including airline tickets, and I was scheduled to leave Tunisia on 17 August for Paris, then fly to New York to connect to Houston, and then travel by car to College Station, where the university is located. I knew the names of my tutor and the main teachers.

One single question was still haunting me:

"Could I return to Tunisia once I had obtained my PhD, and find a suitable job, in accordance with my competencies and within an adequate environment, or would I be an eternal expatriate, obliged to always work abroad?"

I got the answer a few years later.

3.7 Change of Direction (Crossroads – Eric Clapton)

On 24 July, I was in Sousse with my usual group of friends, and we were planning to go and watch the carnival of Aoussou, celebrating Neptune. the God of the sea. Around 1 pm, my father (who had already retired two years earlier from the National Guards) had a serious breathing problem, and we were obliged to take him to hospital. The diagnosis was clear: a serious heart failure. Luckily, he could be treated, and he came back home one week later.

One day, I took him for a short walk, as per the recommendation of the doctor. After returning to our house and before opening the door, he stopped me and asked if I could change my plan of moving to the US. He told me that, in case of a serious problem, it would be easier for me to come to Tunisia from Europe than from the US. I could not disagree with him, but I was reluctant to take such a critical decision on the spot.

I promised him to check what I could do. The first thing I did was to ask my brother and my close friends for their opinion. I had mixed answers. Then, I called Mr Ayrault, my teacher of quantum mechanics at Sup Telecom. He was open and acknowledged that the sudden sickness of my father was an important element to consider. He advised me to decide quickly, so as to find another student to replace me and not lose the position the school secured at Texas A&M, should I decide not to proceed.

Finally, I decided to fly immediately to Paris, where I requested a meeting with the director of the school, Mr Le Francois, to whom I explained my situation. He showed a lot of understanding and, after listening to my explanations regarding my preferred fields of study, he proposed to me a solution which consisted of going to Supaéro, in Toulouse to study aerospace electronics and prepare a PhD in the same field.

Without hesitation, I said yes. Within hours, my destination changed from College Station in Texas to Toulouse in France, and my field of studies from solid-state physics to aerospace electronics.

And so, I was now back to aerospace and aviation.

The choice of aerospace was tremendously favoured by the location in Toulouse and the fact that it would be much easier to find an interesting job in aviation in Tunisia.

After shortened holidays, I flew back, but this time directly to Toulouse. My return to my preferred city was a happy moment, and everything was easy. It was as if I had never left Toulouse.

Within two weeks, I was ready and running full gear. I bought a mini motorcycle called Honda Dax in Europe, I became a member of the rowing team of Supaéro, I joined Air France Aero Club at Montaudran, just 500 m from the school campus and my residence (I was able to walk to the plane and fly), and started rehearsing with a band as a percussionist, before shifting to drums.

I had a failed experience in accompanying a couple of dancers on percussion during a live show at the Daniel Sorano Theatre because I skipped part of the musical score, due to stress and/or simple stupidity.

This harsh experience was a catalyst to push me to have proper musical training to play drums and to learn how to address or perform something in front of an audience, properly and without stress. This was my first step in what is called media training or media coaching.

Thanks to rowing and as part of the training programme, I became a regular long-distance runner. A practice I kept for the rest of my life. This physical activity developed my ability to have a strong will and taught me endurance, patience, and how to overcome pain. The rowing also taught me teamwork and humility: the eight rowers must row at the same pace, and we were never sure to win, even if the competitor had a reputation for being weaker than us.

The flight training was exciting, but very quickly, I knew it would not be my job. I had a fantastic instructor who was a former fighter pilot in the French Air Force, with a long experience in the bombardment of Algerian freedom fighters. He confessed that he was part of the squadron that attacked Sakiet Sidi Youssef in Tunisia in February 1958.

He taught me some tricks about how to attack quickly and pull up in seconds without stalling. He called it the *bombing of the meshtas*, meaning the bombing of the winter shelters, which were hidden in ravines, gorges, and deep valleys in Algeria.

The only memory of this special manoeuvre was that I used to lose any sensation in my left arm, and that I needed to shake my head in order not to feel dizzy. We tried this trick on some castles and remote farms in the countryside. As my instructor was also an assistant in Supaéro, we became very quickly friends, and he helped me a lot in mastering the utilisation of some instruments to control the flight.

One of his most helpful gestures was to introduce me to Mr Klopstein, who was a teacher and a researcher working on the concept of the Head-Up Display (HUD), a system allowing the pilot to have all the essential flight parameters available on the windshield and, therefore, be able to watch outside the cockpit while accessing all the necessary data, particularly during critical flight phases.

Thanks to this introduction, I was able to participate in performing some test flights with the Nord 262 aircraft equipped for the purpose to validate the concept of the HUD. This was an amazing part of my education, and I could grasp the complexity of the test flights. I also had the opportunity to land at different small airports around Toulouse, like Montauban. My participation in this flight test programme had a very important consequence on my future career.

My new encounter with the Nord 262 reminded me of the confusion about its name during my visit to El Aouina Airport in the early 60s.

As for flying, I gave up a few months later, after logging a hefty amount of flight hours. It was the result of a lack of interest and of a lack of time: I had a lot of things on the burner at the same time.

But above all, I was shocked by the death of a French friend, who was a member of our rowing team, in an aircraft accident. He was flying at the Aeroclub of Muret-Lherm, located 25 km south of Toulouse, in a Jodel aircraft, which was a simple, light, single-engine French aircraft.

This accident occurred just a few weeks after I lost a very close Franco-Spanish friend in a motorcycle accident. We were riding several motorcycles on countryside roads, near the Pyrenées, imitating the scenes of the movie *Easy Rider*. The majority of the motorcycles had a passenger, except two or three. Our friend Bruno, who was riding solo, tried to speed up to scout the road ahead of us until we lost sight of him.

After a couple of minutes, we arrived at a crossroad and we discovered his moto under a lorry and himself ejected several metres further, out of the road with his helmet stuck in the pillar of a road panel. He had died on the spot, according to one of our friends who was an internist doctor. It was the first time in my life that I was, physically, very close to a dead person. When my mother died, I was too young, and I could only participate in the ceremonial part of the funerals.

This time, I had to check on the victim, call the police, inform the family, and go to the hospital with the ambulance. It was a real nightmare. His death was so stupid and so unfair that I decided to never buy a motorbike bigger than my small Dax.

During my frequent visits to the Air France hangar, where the Rally and the Cessna aircraft, which were used for the training, were parked close to much larger aircraft like B707 and B727, I used to pass by the office to the chief planner, who was the keeper of the technical status and the regulatory documentation of the two small aircraft.

Each time I passed by this office, I used to have a look at the huge planning displayed on all the surrounding walls, showing all the planned heavy checks up to late 1979. I could distinguish clearly that many slots were dedicated to northern African airlines (which by that time were operating close to 40 B727), which was much more than the whole Air France fleet of the same type.

This fleet was homogeneous in terms of technical specifications and represented a good business opportunity for Air France in terms of maintenance and overhaul. I could also see some North African aircraft in the hangar.

I became, somehow, friendly with the chief planner, and he did not hesitate to explain to me that they needed to plan the activity of the hangars for more than two years in advance, and since the North African fleet was an important part of their business, they planned the aircraft of these airlines by their registration number.

I asked him if the Tunis Air fleet was included, and he promptly answered:

"Of course, we have two or three aircraft up to the end of 1979 and beyond."

I was not surprised. For me, Tunis Air could not master such high-tech activity. The future will show me that this forecast was wrong.

The course programme was very interesting and well-balanced. We had a lot of aerodynamics (by Professor Rebuffet, one of the most eminent aerodynamicists in France), flight mechanics, automatism, electronics, computer science, radio communications and radars, on-board avionics, satellite architecture, launch, stabilisation, and control. I was very interested in aerodynamics and satellite technology.

Due to a close relationship with one of the directors of the CNES (National Centre of Spatial Research), I could access quite a number of publications and attend some conferences of this respectable institution and get updated on various new developments. The proximity of the CNES from Supaéro was a determining factor in pushing me to attend these events.

One of the other regular attendees was my head professor, Leo Thourel, who was my professor of microwaves and radars. This allowed us, after two or three occasions, to better know each other and discuss openly different scientific subjects.

One late evening, in March 1976, after we attended a conference on the preliminary results of the utilisation of the satellite Symphonie, built thanks to Franco-German cooperation, we were walking to the parking, when Mr Thourel stopped me and asked if I would be interested to work on a project of a field-effect-bi-grid transistor to be used as a multiplexer in telecommunication satellites.

In addition, he asked me if I would be interested to be part of his teaching team as of next September. He proposed to me to give lectures on microwaves and radar to the students of the second year who selected the avionics option. One good part of these students come from *Ecole Polytechnique* and were older than me.

I was excited by his offer, but at the same time, I was afraid of the consequences. I asked him how he could offer me all this, while I had not yet finished my current scholar year and had not yet passed all the exams. His answer was very laconic:

"Let us say, I make a bet."

A few days later, I gave my approval.

During the last two months of the school year, I had the opportunity to get my first contact with what was called, then, Airbus Industrie and in particular with its flight test department, its chief Bernard Ziegler, and his deputy Pierre Baud.

Our professor of flight mechanics was working for French Aerospatiale (one of the four founding partners of the G.I.E Airbus Industrie, along with Deutsche Airbus, C.A.S.A of Spain, and initially Hawker Siddeley of UK), he organised a two-day visit to Airbus with a special focus on the assembly line and the flight test department, combined with the nucleus of the new training centre.

During this visit, Captain Ziegler mentioned to our group a small abnormal behaviour of the flight simulator which entered into a slight induced Dutch roll each time the aircraft was ordered to turn left or right, while the real aircraft was not suffering from this problem.

As a challenge, he told us:

"If you can find the origin of the problem, you will have the right to fly the simulator, each one of you for two sessions."

We were five attending. Two answered it was, most probably, a hardware snag and three (including me) a software snag. Captain Ziegler confirmed it was a software problem and that the manufacturer LMT was working on it. They succeeded in reducing the amplitude, but not in suppressing it completely.

He then added:

"I give a chance to those who answered it was a software problem to continue the investigation and I give you up to the end of the week as a deadline."

We asked for a few minutes of internal consultation and came to the conclusion that it was a miswriting of some flight mechanics equation, or simply a missing mathematical term supposed to link the behaviour of the aircraft around its three axes (roll, yaw, and pitch).

One of us who was a computer wizard said to Ziegler:

"I am confident that the core of the problem is that the term of the flight mechanics equations coupling the roll and the yaw is missing."

He explained what the suspicious term was to the Airbus team and to the LMT engineer who was present as field representative and who volunteered to check and come back with the feedback.

On the following Tuesday afternoon, our teacher who had organised the Airbus visit came to the classroom to inform us that our guess was right and that Captain Ziegler invited us for simulator rides in the coming days. It was a rainy Saturday when the three of us arrived at the simulator centre, where we found Captain Ziegler, Captain Baud, and Captain Pinet, the head of Aeroformation, the organisation in charge of training for Aerospatiale and Airbus.

He was the same person who had reached Mach 1 with the Concorde. He was very kind to us; he seemed humble and down to earth. I was impressed by these three captains.

Very quickly, Ziegler took the lead, by explaining the reason for our presence and suggested that each one of us had an introductory session to the simulator (each student with one of the three pilots, my instructor was Bernard Ziegler), followed by another session where we could manoeuvre the simulator as we liked.

After the briefing sessions, we had an opportunity to chat all together and I came to know a little bit more Captain Pinet, which was very helpful for the future.

In June 1976, I obtained my Diploma of Telecommunication Engineer with a specialisation in Aerospace electronics, and was admitted to prepare for a PhD, in the same speciality at the CERT (Toulouse Centre for Studies and Research), which was part of the French national organisation dedicated to military aerospace research: ONERA, and in particular its department called D.E.R.M.O (in charge of microwaves).

Just before travelling to Tunisia, I had to undergo a lengthy and comprehensive interview by military intelligence officers. I was holding a Tunisian passport and therefore they had to check if I did not represent a danger to French national security. This was the first of several such interviews.

I had mixed feelings when obtaining the diploma. On the one hand, I was happy and proud of having achieved such a goal, but at the same time, I wondered if this piece of paper would be enough to allow me to work and bring my contribution to my community, to my country, and to humanity. I felt that I needed to learn more skills and acquire some experience in dealing with people and complex situations. For this purpose, the PHD was an opportunity to achieve this target.

I returned to Tunisia in the summer, with a very important project to realise, namely to get married to my fiancée Ibtissem, whom I had known for some time, not to say since her birth. She is the daughter of an officer of the National Guards who was a close friend of my father. Our two families lived close to each other, in the same city of Sfax for some years.

I used to see Ibtissem regularly, and over the years, our relationship developed from curiosity and friendship to attraction and love. As much as I am interested in science and mathematics, Ibtissem loves arts, architecture, and design. This difference in passions proved to be helpful by creating a positive complementarity between the two of us.

As usual, with the help of my brother and a French friend, Jean Pierre Daubert, we succeeded to pack everything with my in-laws. My father was not in good health conditions and we, therefore, tried not to involve him in the logistics of the wedding.

The ceremony took place on 7 August 1976. On the first week of September, my wife and I were in Toulouse in our house, next to Supaéro. In March 1976, I decided to rent a house jointly with my friend Jean Pierre and leave my room at the residence of Supaéro. Since Jean Pierre left for another city, where he got a job at an EDF power station, in August, I became the sole tenant, which was convenient for me and my wife.

My new life, as part of a couple, did not impact my daily routines because Ibtissem had to follow the courses of a Fashion design institute and she was out the whole day as well. We had nice neighbours and my wife adapted quickly to her Toulouse life. This was a crucial asset for my future career in Toulouse.

A few days after our arrival in Toulouse, I had a new interview with the military security, in my office. The same was repeated two weeks later.

But on the positive side, I started teaching microwaves and radar to the students in the second year of Supaéro, with a specialisation in avionics. I still remember my first day, when I entered the huge amphitheatre and climbed on the platform, in front of the blackboard on which the desk was placed. I looked at the audience, most of the students were my age, and I was among the few having long hair.

Many students were graduates from Polytechnique, and they came directly to the second year of Supaéro, as Ingénieurs de l'Armement, after completing their military training. It means a large part of the audience was from the army and short hair was the rule. I was impressed and had doubts to impose myself and convince my audience. Fortunately, the first lesson was about the basics of microwaves and some reminders about electromagnetism and Maxwell's theory.

It was a domain which I knew quite well, and I did prepare this first lesson very seriously. Professor Thourel came half an hour before the end of the two-hour session to evaluate my performance. It was mainly the Q&A time and, globally, I was able to answer the questions and even give additional explanations.

He eventually joined me on the platform and announced to the students that, due to some frequent travels to Paris and in order to avoid discontinuity, I'd be his replacement for all his classes during the whole month of October. I learnt that at the same time as the students and was surprised. Here I am, going from two sessions, per week to five sessions.

I responded by telling him that I would not have enough time to prepare for the other subjects and to work on my own research programme, to which he answered that I was young, and some sleepless nights would definitely contribute to my training.

These close to 40 lessons I gave, from mid-September to mid-December, were one critical element in shaping my personality and improving my ability to communicate with people and address crowds. I owe a lot to Professor Thourel and his trust in me.

3.8 Another Aerospace Horizon (Interstellar Overdrive — Pink Floyd)

By the end of September, I had a discussion with my direct boss, Mr Priou, about these multiple security checks (including office and document searches), and he promised to do something about it. During this discussion, he also told me that a certain Jean-Pierre Petit, an engineer and researcher at the CNRS, along with another colleague called Maurice Viton of the LAS Marseille, were to come a few days later to conduct some tests on MHD (Magnetohydrodynamic propulsion).

The tests required a magnetic field of 1 Tesla. I volunteered to be part of the team in charge of helping Petit and Viton. After a few days, I met these two persons in the laboratory of the CERT-DERMO. We had a model, in the form of a small plexiglass cylinder, of a diameter of around seven mm and a height of around one cm, positioned on an inclined tray, on which flowed a slightly acidic water.

The cylinder had a sort of electrical wire around it and was put between the jaws of the famous DERMO magnet. I was in charge, along with another colleague, of setting up the cylinder. The whole purpose of this experiment was to see if the setup worked. Jean-Pierre Petit and Maurice Viton had the intention to conduct the real experiment in their laboratory in Marseille.

The test consisted of generating, thanks to the flowing water, the very well-known circles in the water, which are the shock waves generated by the relative movement of an object in a fluid, and then applying both the electrical power and the magnetic field to see their effect on the circles. The result was fantastic. The circles were annihilated.

Simply expressed, a vehicle having a powerful electromagnetic field available on board, perpendicularly to its movement on water, does not confront the famous bow wave (in the air, it would be the shock wave). We finished the tests, and the whole group went for a debrief around some sandwiches and drinks.

It was the first time I heard a serious scientist speaking about UFOs and explaining that, if flying saucers existed, they would be powered by MHD and they would fly at very high speed without any restriction in the air due to the annihilation of shock waves. He also explained that the optimum shape would be a saucer, like what is reported in the media, because it allows to maximise the surface integral.

This encounter with Petit and Viton was an eye-opener on the UFO universe. Since my first readings on Roswell and other UFO events, I was suspicious about their actual existence, but I was always a strong believer in the possibility of the existence of other intelligent beings elsewhere in the universe. In this respect, my Muslim background comforted me in my belief.

The explanation given by Petit during his visit to the CERT (DERMO) in Toulouse, and what I heard during the conference given by Petit and the journalist Jean-Claude Bourret, I believe, in November 1976 in the amphitheatre of Rangueil University of Toulouse, pushed me to dig further into the subject. I consequently read several books (many of the books written by Petit) and other publications.

These readings allowed me to become familiar with the Ummo saga. This was the story of a Spanish group of people, led by a person called Sesma, receiving through dictation or typed documents, detailed information about a given planet, UMMO, distant around 14 light-years from us, and about its inhabitants, their social life, culture, beliefs, and technologies.

The information was diversified and contained a lot of advanced scientific theories and statements. If this was a prank of students or a mystification of a sect, the people behind it must have a very good level of education and have mastered some complex and new theories. Certain mentioned technologies were even unknown on Earth.

To make it simple, the UMMO saga was disruptive in many aspects and involved a lot of people around the world, some of whom were high-level scientists. The global outcome of my contact with Petit and Viton was that I discovered new fields of investigation in cosmology, plasma theory, and even relativity.

I know that some positions taken by Petit, later in his life, vis-à-vis certain events, scientists, and institutions, brought him some critics and even suspicion. Nevertheless, a good part of his scientific work deserves respect and interest.

One very surprising element is that, to my knowledge, up to this day, I never saw any formal, serious denial of Jean-Pierre Petit's works and/or statements on UFOs and some scientific subjects by any recognised and respectable scientific authority.

This parenthesis, opened in the middle of the laborious start of my research, was much welcome. I had wasted a few days but gained some maturity on how to apprehend research. I understood that if you don't have all the necessary elements for an experiment available in your laboratory for the whole duration of the tests, don't dream of meeting any deadline.

In my case, I was borrowing the field-effect bi-grid transistor, manufactured by the Japanese company NEC, from the LEP laboratory (also called the Philips laboratory) in Limeil-Brévannes, next to Paris.

This meant that each time I needed the device, I had to travel by train to Paris and bring the transistor in a special package to Toulouse for a few days and then bring it back to Paris. The reason was the lack of money to buy our own transistor. The idea was to wait until some preliminary concrete results were achieved, before investing.

In addition to this hassle, and the frequent security inspection, I had to face an additional hurdle: my temporary military service exemption expired, and I had to return to Tunis to meet with the military authorities and settle the problem. This was a real blunder generated by the government.

After spending money and wasting our time in the military preparation programme, they cancelled the whole scheme and made us indebted for full normal military service (i.e., one full year). They did not even deduct the time effectively spent.

Most of the students who followed the military preparation scheme, acquired all the basic military skills, like marching in, obstacle course, camouflage, aim and shoot, read a map, and interpret all the relevant elements, and so forth, during their studies. Coming after seven or eight years and asking them to drop their studies, research, or job to join the army was a really stupid idea.

Each one of us had to explain why he could not join the army and agree with the military authorities on a further extension of the exemption, on the option of the civil service where your salary goes to the army, or on the date of the start of the normal service. These discussions used to take a few days or a few weeks, depending on your marital status and the level and complexity of your studies.

3.9 Get Back to Where You Once Belonged (The Beatles)

I went to Tunisia around December 20, one day after another security inspection, which was close to making me lose my temper due to the nature of the questions. That day, I seriously considered definitively stopping my work at the CERT-DERMO.

Initially, the purpose of my trip was to sort out my situation vis-à-vis the military service. I was naively thinking it will take a few days and that I could be back to Toulouse by the year-end.

Unfortunately, despite being polite and patient, the simple fact I said I did not understand why we must repeat again all that we had already done during the preparation cost me a delayed appointment on January 10. In the end, I got a simple extension of the exemption for one single year, starting around the end of January.

Meanwhile, several critical developments took place. First, by mid-November, the husband of one of my cousins, Ali Haddad, who worked as a senior engineer in the Tunis Air technical directorate, sent me a letter telling me that the airline was planning to hire additional engineers of my profile.

One week later, I met, in the streets of Toulouse, a senior captain of Tunis Air (Moncef Chebbi), with one of my Tunisian friends living in Toulouse.

After being introduced to the captain, he asked me if I had an experience with aircraft simulators. I told him about my encounter with Captain Ziegler and my few days at Airbus and Aeroformation. He immediately asked me to come and join Tunis Air to take care of the project related to the acquisition of a B727 simulator.

I told him about the information I got from Ali Haddad, and he confirmed that, true, the recruitment would be done by the Technical Department. I also told him that, due to some complications with the military authorities, I would surely have the obligation to fly, soon, to Tunis and that I'd come to see him. He gave me his business card, and we left it there.

Second, one day after my arrival in Tunis, I had a desperate call from the DERMO about an urgent meeting with some people from the DGA (Direction Générale de l'Armement) about my budget and the need to send a new amended version, following a rejection of my first draft.

This meant that I would have to accept to continue working with a borrowed transistor, at least, until July 1977. At the same time, I was requested to go to Paris to obtain a security clearance from the DGA. I started thinking that Mr Thourel made a mistake when he offered me the position because he did not suspect all the implied complications.

In view of the two simultaneous developments, I decided, on 24 December 1976, to stop my adventure in Toulouse and come back to Tunis. I spent the last few days of the year and a good part of the month of January, organising my move.

With great hindsight, I am convinced that the day I decided to return to Tunisia, I gave another direction to my career and to my whole life. It was a critical turning point which defined my future and a breaking point which estranged me from many friends in France.

The love of my country and the desire to serve Tunisia and to come back to my roots were very strong and generated a great amount of happiness. But my passion for an exciting job and a successful career was making me feel worried about the risk of any bad surprises. Ultimately, I made a bet on Tunis Air.

4. Learning the Aviation Job in Tunisia (Night in Tunisia-Dizzy Gillispie)

The sudden return to Tunisia generated a certain number of problems, including upsetting my best friend, Michel Forgue, whose wedding I could not attend nor be his witness. Despite a long letter of explanation, he never accepted my apology, and time made its job by separating us. This was my saddest collateral damage.

Another problem was the wrong understanding of Mr Priou, who believed that I dropped because I did not know how to structure and conduct my research. At least for this case, I had the opportunity, a few months later, to meet him in Toulouse and correct his misunderstanding. I hope I succeeded.

Another problem was my feeling of guilt for not taking enough time to explain the reasons for my sudden departure to my teacher and mentor, Professor Thourel, whom I count among the handful of people who tremendously influenced my career.

The sudden return to Tunis was not easy, despite the fact that my wife was happy to reunite with her family. We had no house, no car, no furniture, and little money to spend because I had not yet started receiving the monthly instalments of my student/researcher scholarship, and I just had received half of the salary for the teacher's job. My in-laws offered to give us a room until we organised ourselves and found our own apartment.

4.1 Landing at Tunis Air (4&20 Years Ago-Stephen Stills)

On 2 January 1977, I started my new job as an engineer in charge of technical studies at the Technical Department of Tunis Air, located at El Aouina, next to the airbase and the military barracks where I had done my military preparation a few years earlier. The place was familiar, and I received a warm welcome from the people I met on the first day.

Simultaneously, I was going two or three times a week to the military headquarters to sort out my military service problem (it took a few weeks).

I was very happy to get this job for several good reasons.

First, I was able to serve my country in a position which could enable me to work in a high-tech environment.

Second, it was a position which allowed me to enjoy one of my favourite passions: aviation.

Third, the team was young and enthusiastic.

And last but not least, the boss was a charismatic, self-made man who loved aircraft and who was leading by example. He did not hesitate to recruit many engineers who were much more educated than him and empowered them to work freely within the frame of the set rules and regulations, without interfering, while ensuring efficient supervision in an intelligent way.

This man, Mohamed Miaoui, is one of the handful of persons who had an important impact on my career. I was coached and supervised by the manager of the studies division, Mr Ammar Trabelsi, a graduate from Supélec Engineering School in Paris, who taught me a lot of concepts related to feasibility studies, risk assessment, and other economic tools not usually within the field of expertise of an engineer.

He was clever and kind enough to allow me to interact directly with the big boss and to spend part of my time with the maintenance crews in the field, learning the real aircraft troubleshooting with a hands-on approach.

I had several projects to study at the same time: acquisition of a B727 flight simulator, the installation of a long-range communication (high-frequency or HF) network, the installation of a short-range communication (very high-frequency or VHF) network, the implementation of a computerised reliability follow-up of all major components with the aim to define the optimum spares inventory, the computerisation of the spares inventory, etc.

After a few months, the simulator project was shelved for lack of funds, and I had free hands to accelerate all the others. At the same time, I agreed with my director to get practical on-the-job training as an avionics technician, both on the Caravelle and on the B727, which became the workhorse of the Tunis Air fleet with 10 units by mid-1977, and to have a B727 type rating as avionics technician.

From the first few days at Tunis Air, I had to deal with the political landscape as it was at that time in the country. First, I discovered that most of the trade union officials were also militants of the Destour Party, which was the only party in the country, controlling all the levers of power. To avoid any problems, the name of the game was to be at least a member of the party.

For the senior engineers and managers, it was not mandatory to be a member of the trade union. It took the group in charge of the activities of the party and the trade union a few weeks before they first approached me. It was by a simple invitation for a cocktail celebrating the re-election of some people at leading positions in the party cell within Tunis Air.

I did not attend. Another day, I found a membership card of the party on my desk with simply my name on it. I shredded it and ignored completely what happened. Again, no visible reaction for some time. But one of the old-timers told me that I was taking some risks and that I should be careful.

4.2 Airline Daily Life (Sultan of Swings-Dire Straits)

My first year at the Technical Directorate of Tunis Air was very fruitful. I learnt how to troubleshoot and repair any avionics problem on an aircraft. I finalised the HF and VHF projects and started the installation and the deployment of both networks, which were fully operational, respectively, in 1978 and 1979.

We launched the project of automation of the spare inventory management at the same time as the optimisation of the inventory, thanks to a precise calculation of the reliability of the most valuable components. It was a real pleasure to work with the technicians, the team leaders, the controllers, and the foreman to troubleshoot an aircraft during a transit stop of 45 minutes under a temperature of 40° Celsius in the midst of the Tunisian summer.

It was a sort of dual physical and intellectual challenge consisting of finding a problem and solving it, while reading manuals and pilot reports, doing some tests and avoiding spilling sweat on the documents or getting wet hands, in less than 20 minutes.

These moments were, for me, the real opportunity to learn what real airline operation were and to first apprehend the complexity of an aircraft. For three years, I succeeded in having several periods in a working week to spend with the teams in the hangar or in the line maintenance, troubleshooting and repairing avionics systems.

I really owe a lot to these joyful, competent, and friendly people who, despite modest salaries and harsh living conditions, were doing a fantastic job. For me, they were on par with their colleagues in Europe or the US. In some respects, they were even better because they did not have the same facilities and the same comfort at work.

Another group of technicians got my full respect: the mechanics specialised in structural repair.

They helped me install all the antennas for the communication networks quickly and efficiently in the heat of the summer, inventing clever solutions not even envisaged by the manufacturer to stabilise the antennas without adding weight.

During these installations, I appreciated the spirit of camaraderie, team effort, and desire for perfection. Everybody participated in the group efforts, whatever his position or qualification. The team leader and the warehouseman worked side by side.

My practical on-the-job training allowed me also to learn the language of aviation with its own expressions and codes. I had already learnt a lot at Supaéro, during my flight training, and thanks to my readings for several years, but it was at Tunis Air where I got a structured approach. Like any other field of activity, aviation has its own expressions and codes.

From ATA chapter to AOG, passing by deadheading, go-show, and others. But the beauty of aviation is that airlines have their own communication network called SITA, which they co-own through a cooperative structure, and I had the honour and the pleasure, during close to four years, to be the representative of Tunis Air at SITA and a board member of this cooperative.

Airlines used to communicate between themselves and with the manufacturers, the aviation service providers, airports, and civil aviation authorities through telexes. In addition, aviation uses the phonetic code of NATO to designate the letters of the Latin alphabet.

In summary, airlines have a completely self-contained and self-sufficient environment, where they use their *own language*, their *own alphabet,* and their own communication network.

The HF communication system was very innovative, and we contributed to its design. It was based on scanning the most probable or optimum HF frequencies, a function of the sunlight during each period of the day and period of the year, according to the frequencies chart published regularly by Bracknell Centre in the UK.

Thanks to this technique, some major events in the life of Tunis Air were either detected in advance or followed closely, which allowed me to be a direct witness of these events. Some of them made history.

While I was busy with my different projects, I was asked by my director to help the team, comprising people from different directorates, negotiate the lease of a B737 from a company called ILFC, based in Los Angeles. It was the first time I knew about the existence of leasing companies and got acquainted with Steve Udvar Hazy and his team. Steve later became a good friend.

In 1978, I was appointed head of the Components Overhaul Department and was asked among the other colleagues, to repatriate in-house the biggest possible part of the maintenance of the Tunis Air fleet, including the structural heavy maintenance (D-check), if it made sense economically.

During the same year, I enjoyed the birth of my son, Karim, on 13 September. Due to business obligations, I had to travel the third day after his birth and leave him with his mother, who was still recovering in the clinic. It was a very tough decision, but it was not the last of that kind during my career.

The trip to Cincinnati, Ohio, in the US was mandatory because I had to attend an important event related to non-destructive testing (NDT), in the frame of the preparation of the upcoming D-check or overhaul of our first B727 in-house. NDT was an important element of the D-check because, thanks to X-ray, Doppler, or Magnaflux technology, we were able to detect cracks, corrosion, or other material problems on the aircraft and its components and repair them.

I discovered that an aircraft, like a human body, could be checked by the same techniques. General Electric (GE) was sponsoring a conference, organised by the Air Transport Association (ATA), to present the new techniques and methods of investigation by NDT. It was a must to be up-to-date in order to ensure maximum efficiency during the D-check.

Over the next three years, we increased the in-house share of the fleet maintenance to the point we achieved the first D-check of a B727 in Africa and the Arab world by early 1979, in the record time of one month, thus contradicting the forecast of Air France.

After a few months, I went with my boss to the Air France Maintenance Directorate in Orly, France, to close some open items in the cooperation agreement. They openly told us that they kept the slots planned in 1979 and 1980 in their planning until two days before the test flight of the aircraft, which has undergone a D-check in Tunis. It must be underlined that the test flight was immaculate, without a single snag.

Boeing, which was following closely the performance of the check, was briefed daily by its field representative and sent one inspector to help, just in case. This inspector was a fan of Steppenwolf, like me. Therefore, I invited him to my house for dinner and to listen to our favourite group.

During this dinner, he openly told me that there was a contingency plan in Boeing, in full consultation with Air France, to send a sort of recovery team to help in case things did not go well. However, the field representative sent consistent SITA messages, saying that all was going well and that everything was under control. Therefore, there was no need to prepare a contingency plan.

The field representative was a US citizen of French origin, a former Tomcat pilot, and a brilliant engineer who had shown strong support for our company and stood by our side at all difficult moments. His name was Robert (Bob) Verdier.

One of the things which impressed him most during this check was how we succeeded to machine a main landing gear ball joint, made of brass, with a simple lathe during a night shift with the technician being fed spicy Tunisian sandwiches full of harissa and tuna, and provided with regular sips of heavy stuff. The ball joint was not available for purchase at the landing gear manufacturer, nor at Boeing, Air France, Sabena, or other neighbouring airlines.

We had to follow the motto of our national hero, Hannibal: *either we find a way or we build it*. Since I oversaw components maintenance and overhaul, we had to build it.

For some reason, it happened that we had a cube of brass with dimensions slightly bigger than those of the ball joint, which could be used to manufacture the critical piece. We found the right technician and the right machine, and we decided to go ahead. It was the only way to proceed.

Otherwise, the landing gear would not be ready for installation on the aircraft on time for the test flight.

We started at 4 pm, the same day, with a small team including the technician, an assistant, a guy from the engineering with all the necessary documents, and myself coordinating. We have prepared enough Tunisian sandwiches of different sorts and different sorts of drinks to help us stay awake during the night ahead.

Finally, my real role was to feed the team and serve the drinks, which proved to be more complicated because of the presence of a small crowd watching the work under process. Many people, including some managers, were so curious to see by themselves if the challenge could be met that they had decided to stay and watch.

Everything went well. By 4 am, the ball joint was ready, fully shining under the workshop lights.

This commando action, avoided us a delay and saved us money. The director, who came at around 5 am, did not stop laughing when I told him about the sandwiches and the drinks. He decided to pay the drinks from his own pocket as a sign of appreciation.

He commented:

"We need to properly celebrate this achievement, after the test flight."

The Boeing field representative arrived at around 7 am with a lot of croissants and other pieces of pastry and rushed directly to the workshop, curious to discover the result. After a short briefing, he cross-checked with our chief controller the dimensions of the ball joint. He was impressed by the rigorous precision of the ball and its high-quality finish. As an additional sign of appreciation, he returned home and brought back more drinks.

In total, the feat of our technician was celebrated by the whole maintenance department. Since it was early morning, everybody in the workshop jumped on coffee and croissants, leaving the other drinks for later. All those who were present during the night shift did not go home, because they were keen to watch if the ball joint would fit perfectly in the landing gear.

Around midday, the landing gear was fully equipped and ready to be installed on the aircraft. The same night, the ground tests started, according to the pre-set planning.

Following this success, the B727 D-check became a routine maintenance operation in Tunis Air.

Despite a lot of efforts and many signs of goodwill towards our friends at Royal Air Maroc, they never accepted to send us their B727 for D-checks. Over time, I discovered that there was no real willingness to cooperate between the four North African airlines.

It was a question of national pride, jealousy, and, almost importantly, a priority to create jobs in the country and save hard currency. There was no real long-term vision to share the workload and maximise benefits. Every country claimed brotherhood and lateral cooperation, but everything was directed south/north. Brotherhood was in the words, not in the actions.

During my nine years at Tunis Air, I discovered that we had no relations with Libyan Arab Airlines. We had limited relations with Air Algerie (Tunis Air made a gesture by sending B737 pilots for training on the Air Algerie simulator), friendly relations with Royal Air Maroc (RAM), but it was a one-way ticket.

Tunis Air gave RAM the maintenance of some avionic components, was the first to use the RAM B727 simulator, and later added the RAM B737 simulator for pilot training, without RAM reciprocating at all.

But, for the sake of honesty vis-à-vis a man who was the most prominent person in RAM and an important personality of the Kingdom of Morocco, I must report the noble act of late General El Kabbaj.

In 1983, I was on duty during a weekend and received the news that one of our B737s had suffered a double engine FOD (Foreign Object Damage: in the present case, the impact of a flock of birds) during take-off from Toulouse airport.

This meant that we had to send two engines to Toulouse, recover the two damaged ones, and send them to Brussels to be repaired by the Sabena Technical Department.

My problem was to find, very quickly, a cargo aircraft available to accomplish the mission: fly to Tunis, take two engines, fly to Toulouse, offload the two engines, load the two damaged engines, fly to Brussels, and offload the two damaged engines there.

I tried to locate an aircraft for half a day, but it was mission impossible. There was a remote possibility with a cargo charter airline, but both the timing and the cost were not acceptable.

Before giving up, I called my good friend Mr Azzaoui, director of Ground Operation at RAM, to ask him if he could help me find a cargo aircraft.

Half an hour later, he called me back with the good news that he had located an aircraft and that the details would be given to me by General Mohamed El Kabbaj himself, RAM Flight Operation Director and Chief of Staff of the Moroccan Air Force.

I was intrigued by the involvement of this general. I knew him because I had met him a few times to sign the training agreement and the renewals, as well as during some conferences.

He was grateful for the fact that Tunis Air used RAM simulators.

He was always teasing me by asking me how it was that the Tunisian pilots accepted somebody from the *ground* to be their boss, always adding:

"To be fair with you, the pilots are not the best managers."

In 1981, 1982, and early 1983, I held the position of Flight Operations and Ground Operations Director. A few minutes later, the General called me to tell me that a Hercules of the Royal Moroccan Air Force would depart in a few minutes from Rabat to come to Tunis, take the two engines, carry them to Toulouse, and subsequently transport the damaged engines from Toulouse to Brussels.

I thanked him several times for providing us with the so badly needed airlift and asked him how we could settle the cost of the charter.

His answer is unforgettable:

"In the army, we don't sell anything, we can only give our blood. Knowing the Tunisian hospitality, I don't even ask you to feed the crew—I know you will do it."

Everything happened as planned. We invited the crew for dinner and covered the handling cost at Tunis Carthage, as well as the fuel cost.

I believe the general tried to reciprocate by this gesture of a lord (Seigneur). He was indeed a real *Seigneur*.

I will never forget how he saved King Hassan II when he was the captain of the B727 transporting the king from Paris to Rabat on 16 August 1972.

The flight was attacked mid-air by F5 jet fighters from the Moroccan Air Force in a tentative of *coup d'état* organised by General Oufkir, known as *the Airmen Coup* (le coup d'état des aviateurs).

Thanks to his pilot skills, his experience as an ex-fighter pilot, and his intelligence, he succeeded in landing safely with a severely damaged aircraft (two engines, the rudder, and the stabiliser were damaged, plus several bullet impacts on the fuselage) and saved the life of the king. Since then, he became one of the most important persons in the kingdom.

I had the opportunity to meet him again after his gracious gesture vis-à-vis Tunis Air, years later, when I moved to another company. The last time I had contact with him, I was in Toulouse, and he called me to ask for help to get an urgent appointment with a famous doctor. I succeeded in getting the appointment, but I then received a call from one of the top managers of Royal Air Maroc telling me that he would not be able not make it. He passed away a few days later.

Unfortunately, during my career, I could see the same lack of long-term vision very often, in many countries. It seems that there are very few politicians who have real visions for their countries and work to achieve them. The majority have slogans and short-term targets.

I consider myself lucky for having shared a good part of the '50s, '60s, and part of the '70s with a collection of great men who have shaped this world.

All of them were visionaries, dedicated to their respective countries, with strong leadership qualities and a lot of charisma. Unfortunately, some of them developed dictatorial behaviours and caused some damage to their countries, but a large majority succeeded to make their countries a better place to live in. Bourguiba of Tunisia was one of them, but he did not know how to retire on time and with dignity.

His contribution to Tunisia is largely positive, but some of his late-life errors are still affecting the development of the country. In fact, he was by far one of the very few in Africa and the Arab world who had a real vision for his country and favoured a fully integrated development with a focus on the human factor.

Despite the limited resources of the country, he succeeded to achieve what much richer countries could not do: a united, peaceful people, proud of their country while being open and tolerant.

This was the Tunisia I was living in in the late '70s. Some clouds were gathering, but the country was in a relatively good shape.

Things started degrading from early 1978, and it took some time—up to 7 November 1987—to put an end to the Bourguiba era through a *medical coup d'état,* which saw the then-prime minister, Zine El Abidine Ben Ali, call a group of doctors to declare President Bourguiba senile and unable to fulfil his duties as head of the state.

I leave it to political analysts to explain what happened exactly and the reasons behind it. For me, I had a few more years of exciting times and challenges at Tunis Air before things started degrading.

The first major event was the riots of 26 January 1978, which saw Tunisian forces killing Tunisian people and the unique trade union organisation militants fighting against the unique political party militants after more than two decades during which the two organisations were working together very closely—to the point that many people were wearing the two hats.

The second event was the hijacking of a Tunis Air B727 operating a domestic flight on 14 January 1979, the day we decided with my wife to take my son, who was four months old, to undergo a small surgery suggested by his doctor. The appointment with the clinic was at 8:30 am, but I had been following the hijacked aircraft for several hours already.

I could free myself, literally, for less than one hour to ensure that my son was okay before going back to the operation centre, where I was monitoring all the messages received via the HF station—a valuable tool to obtain reliable information.

The captain had just tuned in the adequate frequency and continued speaking normally with the co-pilot or the hijackers while simply pushing the microphone button.

There was no answer from our side, of course—it was a silent room. But from the ops room, we could hear what was going on in the cockpit. Thanks to this trick, we could have confirmation that they landed first in Malta and then in Tripoli, listening to the requests of the chief of the commando, who tried to be hard and nasty through harsh words and a loud voice.

All the terrorists in the world have the same pattern of behaviour, as if there are standard courses to train people to become terrorists. Shortly after, the captain was summoned by the commando to communicate the list of the requests to the Tunisian authorities through the usual frequencies used to communicate with the flight controllers of Tripoli airport.

There were other channels of communication, but we were acting as a monitoring centre to double-check. Very quickly, we understood that Libya was behind the operation, which ended peacefully. One thing I learnt from this event was that cold blood was the recipe of success, and the captain of the aircraft was a master in this game. His name is Captain Arfaoui, and I have the greatest respect for him.

During these first years at Tunis Air, I had a fantastic opportunity to learn the real world of aviation and understand its constraints and challenges. One thing was sure: each day was a new source of surprises, challenges, disappointments, and sometimes (not often) satisfaction.

In 1979, Tunis Air also saw the arrival of the first two B737-200 in the fleet, with a two-man crew cockpit. From that date, the question of the future of the flight engineers, after the retirement of the B727s, was put on the table and was openly debated. From there on, any decision about the future fleet had to take into consideration how to solve the problem of the flight engineers.

As far as I can remember, there was no noticeable opposition to the introduction of the B737 with a two-man crew, besides very limited negative comments from a group of pilots and flight engineers who were known to be cacique and prompt to criticise anything new.

The same year also saw the unfortunate accidents of two B727s taking place the same day, within two hours of each other—one in Jeddah and one in Tunis. Only the Jeddah accident caused a fatality: the death of one worker on the runway during an aborted take-off, with all the passengers and the crew evacuating safely. The Tunis accident occurred during the landing of a ferry flight (used for training), with no passengers on board.

The two accidents caused serious damage to the aircraft, which needed very heavy repair. On this unfortunate occasion, I had the opportunity to appreciate how the Boeing customer support organisation worked. I was among the team who contacted Boeing to inform them about the accidents.

The first time we called them for the Tunis case, they asked us about the type of aircraft, if there were casualties, the location, the conditions of the aircraft, the position of the aircraft on the ground, etc. Then they offered to send a recovery kit and an assessment team. When we called them the second time about the Jeddah case two hours later, they thought we were repeating the same story.

It took us few minutes to explain that it was another accident. Here again, they decided to send a recovery kit and an assessment team. The Boeing organisation was capable to deal simultaneously with two emergencies, without any hesitation, and could dispatch all the necessary human and technical resources in a very short period of time.

I followed the recovery of the two aircraft, their repair, and their return to service. It was impressive to see the efficiency of the customer support organisation of Boeing. This dual exercise made me understand the vital importance of after-sales support and how the choice of an aircraft must not only be based on performance and price.

Among the best memories of my interactions with Boeing, I must mention my first encounter with a Native American. One of the key players at Boeing delivery centre was a Native American. He was among the most efficient and competent people in the Boeing delivery team. He was, at the same time, friendly and had a great sense of humour.

One day, the technical director of Tunis Air (my big boss) was in Seattle and, for some reason, asked one of the Tunis Air managers if the Native American was American.

The manager answered:

"Let us ask him," knowing very well the answer.

The reaction of the Native American was explosive:

"What the hell is this stupid question? If I am not American, who then is American?"

Thanks to some exchanges with the Native American, I could learn more about the native nations, their sizes, and their present living conditions when they chose not to integrate the cosmopolitan population of the US.

I came also to know about Native American craftsmanship and, in particular, the sand painting of the Navajo and their jewellery, of which I bought few items. After so many years of intoxication by the movies about the West conquest—with the good white man and the villain Native American—it was a real eye-opener.

After this first Native American, I met others from different tribes or nations and at different positions in society. Some behave like immigrants and speak about the present much more than the past, and others are still proud of their origins and like to speak about their ancestors and their culture. This is America.

The visits to Boeing were also a good opportunity to meet our good friend Mohamed Abdelbari, the permanent representative of Air Algérie in Seattle. He was so kind that he was offering us to use his office (Tunis Air did not have a permanent office, because the policy was to send a representative only during the last few months of the manufacturing of the aircraft).

Mohamed was a funny and entertaining person, coupled with a very competent engineer. I have memorable souvenirs with him, and we stayed close friends, even after he moved to Boeing. Our paths crossed several times, during many years.

Our common joke was:

"We are two parallel lines which cross from time to time."

4.3 The Airbus A300-B4 FFCC (The Two-Man Crew Saga) (Satisfaction-The Rolling Stones)

After, the D-check experience and the satisfactory installation of the two communication networks, I was appointed member of the committee in charge of evaluating the future wide-body aircraft to be purchased by Tunis Air. The choice was very limited and somehow tricky.

If the acquisition were to take place before 1980, the only options offered were the Airbus A300B4, the McDonnell Douglas DC10, and the Lockheed L1011. But if the purchase was to be after 1982, the A310 and the B767 could be added to the list.

Very quickly, we decided to evaluate the A300B4 (and the A310 at a later stage) and the B767.

As of 1979, the committee started travelling to Toulouse and Seattle, collecting information and all the necessary data to carry out a thorough evaluation. At the same time, visits were organised to Air France, Lufthansa, Swissair, KLM, United Airlines, Continental Airlines, and Northwest to get their views on the different available products.

As I was familiar with Toulouse, I was dispatched to Airbus, while another colleague who was familiar with Seattle was dispatched to Boeing. The visits to the different airlines were very often organised during relatively short transits through the home bases of the concerned airlines.

One important element was the good preparation for the different visits. It was like preparing for an examination: read all important information about the company and the products, and prepare the list of relevant questions.

During our visit to Airbus, we were greeted by the team which used to come to Tunis: the salesman, the marketing manager, and the technical specification specialist.

To our big surprise, on the second day of the visit, we were received by the three top managers of the company: Roger Béteille, the General Manager; Felix Kracht, the Senior Vice President of Production; and Bernard Ziegler, the Senior Vice President of Flight Tests.

Knowing that Béteille and Kracht were among the founding fathers of Airbus Industrie, the joint European company (G.I.E: Groupement d'Intérêt Économique), I decided to ask them a few questions about the company. Both were keen to answer without any hesitation and took all the necessary time to describe everything.

I learnt a lot about how Airbus was created and the pivotal role played by Franz Josef Strauss, the Minister-President of Bavaria, who had been instrumental in securing the necessary funding for the creation of Airbus Industrie on the German side. I had already gotten a good idea about Airbus Industrie during my time at Supaéro and my encounter with Bernard Ziegler, Pierre Baud, and Jean Pinet, and I was intrigued by the concept of the G.I. E.

Therefore, I asked Mr Béteille to give us some precisions. He accepted and explained in a simple way that Airbus Industrie (as it was called for a long time) is a pure coordination structure, in charge of fronting the customers in terms of sales, marketing, support, and understanding of their needs, collecting the funds from the customer at delivery of the aircraft, and ensuring coherence between the four partners or founding *fathers* (French Aerospatiale, German Deutsche Airbus GmbH, Spanish CASA, and UK BAE—the names evolved over time) in the fields of engineering and production.

The GIE was also the sole interface with any other external entity, such as, for example, the certification authorities.

Mr Béteille then added a comment, which I did not quite understand at the time but did later on when Airbus became an integrated structure:

"The beauty of this setup is that each partner is responsible for its mistakes and delays and has to compensate Airbus Industrie for any shortfall. (This concept was clearly demonstrated when BAE had to pay hefty amounts of money to Airbus Industrie as compensation for the long strike of the British workers)."

Roger Béteille seemed satisfied with the setup. He tried to give the Tunis Air team a complete picture of Airbus, its organisation, and its people. He then asked Felix Kracht to give us an overview of the industrial setup of Airbus Industrie, and Bernard Ziegler to give us a complete review of the aircraft specifications and performance.

An element surprised me. Roger Béteille was always wearing a white tie on a white shirt. I did not dare ask him the question why. It was only in 1985, one month before he retired, that he gave me a hint, without me even having raised the subject. He had invited me for lunch at the Airbus restaurant, in the guests lounge, and we were discussing the fate of some prominent aviation personalities.

Commenting on the end of life of one of them, he sadly nodded and said:

"His family never accepted his death,"

Then he looked at me and added:

"It happens quite often that we don't forget our dead, but we must never refuse the reality of their departure. The only thing we can do is to keep, through discrete symbols, their souvenir alive."

He did not elaborate more, nor I did insist, but I got a strong impression that this white colour could be linked to a souvenir of somebody. Later on, I got the full explanation, which is very sad and that I cannot disclose it out of respect for the privacy of this great man and his family.

In fact, I still feel very honoured by the trust Mr Béteille demonstrated to me with this comment because his usual answer to the many people who asked him the same question was:

"This is the best-kept secret in the industry," adding,

"And also, people then gossip about my white tie and not about something else."

I became a regular visitor of Mr Béteille from our first encounter in 1979, and progressively our discussions became more diversified. I was literally honoured to exchange openly and frankly with a man who was close to the double of my age, with such achievements in his career.

One of his two secretaries, the late Caroline Huisman, was surprised by our proximity and used to tell me each time I came to see him:

"You are lucky, he likes you."

Caroline became one of my closest friends until the end of her life. She was one of the main pillars of Airbus. She worked with Béteille, Pierson, and Forgeard and was sort of a living memory of the company. I'll never forget her.

Roger Béteille and Felix Kracht were, with Franz Josef Strauss and Henri Ziegler (father of Bernard Ziegler), the former president of Sud Aviation and then president of the newly created company S.N.I.A.S or Aerospatiale (resulting from the merger of Sud Aviation and Nord Aviation in 1970), the founding fathers of Airbus. Each one played a role, but Béteille was the central piece, closely helped by Felix Kracht and Franz Josef Strauss.

Both Roger Béteille and Felix Kracht were seconded to Airbus Industrie G.I.E (Grouping of Economic Interest), by their respective companies (Aerospatiale and VFW Fokker—a member of the Deutsche GmbH grouping) because of their competencies, their deep European convictions, and their strong complementarity. Both of them were multilingual (French, German, and English).

Roger Béteille was the long-term planner, the cold organiser, and the meticulous manager, while Felix Kracht was the very outspoken, hands-on organiser, the crisis manager, and the Mr Fix-it.

Felix Kracht, who was the alter ego of Roger Béteille, played a big role in making Airbus a creative, agile, and efficient company.

Based on his previous experience with other collaborative aircraft programmes, such as the military transport C-160 Transall, and having also observed the problems resulting from the Concorde industrial organisation (each country had its own final assembly line and flight test centre, for example), he defined the complete industrial process in a manner that avoided any duplication of task, while also taking into account the competencies of the respective partners.

This was to make it the most sensible and efficient industrial setup. Each partner was tasked with—and fully responsible for—a complete section of the aircraft and would deliver that complete section, already fully equipped with all the wirings and other equipment, to the final assembly line in Toulouse, where all the sections would be integrated.

As a result of this work-sharing between France, Germany, the UK, and Spain, the amount of work performed in Toulouse was a mere four per cent of the total production work. While Toulouse was the most logical location for the final assembly line (already existing flight test centre and equipment, a very long runway, and, above all, free airspace that allowed unhindered flight testing), Hamburg was put in charge of the cabin furnishing.

In broad terms, this setup still prevails today. The only major changes are the additional final assembly lines, first in Hamburg and later in China and the US.

Felix Kracht was the one who defined the tolerances of all the machines in all the factories (tolerances of less than a quarter of a millimetre) to ensure compatibility during the assembly: it must be as easy as a Lego game. For this, he had to deal with the European measurement units and the English and American measurement units.

To overcome this hurdle, he put each partner in charge of ensuring that the joining of their section with another one was going to work, hereby making each *responsible* not only to produce their section but also for ensuring that it would fit to another one.

Therefore, each of the partners had master moulds of the other section on which their part was to be fixed, and, before shipping his section to Toulouse, ensured that it would perfectly fit on that master mould.

Felix was an interesting character, contrasting with Béteille. He was an extrovert, enjoying life with all its good things, and a heavy smoker of basic French Gitane cigarettes.

He was the opposite of Roger Béteille, who was introverted, sober, and forbidding smoking in his office and around him. The great joke was to see Felix sitting in Béteille's office below the *No Smoking* sign, placed in a strategic location in the office, and heftily smoking.

Very often, Felix used to make his sarcastic comment, "Refrain from smoking. Where does it say?"

In some ways, Felix was more French than German, speaking French fluently without the Toulouse accent. He looked much more like a French farmer, especially when speaking with his cigarette in his mouth, than a German engineer.

This was surprising but logical because he had worked in France from 1947 to 1968 (with Nord Aviation) before joining the future German partner of Deutsche Airbus, VFW Fokker, to work on the Airbus programme as a representative of the German side. He became even critical of the German mindset.

Each time the Germans showed some hesitations in making a decision—and this happened quite often—he used to say his famous expression: *Das sind alle Bundesbedenkensordentraeger*, which means, *They are all wearers of the federal order of procrastination.*

I had the chance to be invited by Felix Kracht for lunch, during which he explained to me the industrial organisation of Airbus and its partners. I was alone because the other Tunis Air colleagues were split between customer support and flight tests. I had several opportunities to meet with Felix Kracht after that, and each time, it was an enjoyable moment.

During the third visit, Mr Béteille organised a large meeting in his office, including Captain Bernard Ziegler and Felix Kracht. From the Tunis Air side, we were four people, including my colleague Hamda Hajji, the head of engineering, who was my partner in most of my assignments abroad, the chief pilot, and the chief flight engineer. Just at the beginning of the meeting, Mr Béteille threw a bombshell.

"I understand that you are asking a lot of questions about the A310 and the two-man crew cockpit."

This was true because, during our second visit, we had learnt that Airbus had launched a detailed engineering study, following a discussion with Garuda Indonesia, to modify the cockpit of the A300B4 and make it compatible with a two-man crew operation, which means the suppression of the position of the flight engineer.

It must be underlined that, at that time, the A310 was designed to be operated by a crew of two persons only, but some airlines, like Air France, insisted to keep the three-man crew cockpit configuration to avoid problems with the pilot unions, which were invoking heavy workload and safety issues.

Here, it is worth mentioning that the B767, which was launched around the same time as the A310, simultaneously with the B757, had a slightly different evolution in terms of crew composition, as explained further down, because of a strong and bizarre opposition of the pilot unions.

Without waiting for any further comment from our side, Mr Béteille added:

"I have an elegant and cost-effective solution for you. You order the A300B4, and we will modify the cockpit for you to become the A300B4 FFCC (Forward Facing Crew Cockpit), as we promised to do it for Garuda, for a really limited amount of money.

"In addition to what has been proposed for Garuda, we could also install a fully ARINC 700-compatible avionic suite (digital equipment) to replace the old analogue suite. We could not do it for Garuda due to time pressure and lack of interest from the airline. In case you agree on this solution, you would have a solid and reliable aircraft with a new cockpit and new digital technology, which is much more efficient and reliable while consuming less energy."

It took us a few minutes to grasp the full extent of the proposal and evaluate its cost-effectiveness. It was true that the cost of the modification was very modest, but we had to compare it to the other available options on the market.

The meeting lasted for more than two hours, and a lot of questions were asked, including how Airbus could train Tunis Air crews for a two-man operation and certify the aircraft in France while all the civil aviation system in the country was controlled by unions opposed to the two-man crew cockpit. After the meeting, we were taken to visit a makeshift mock-up of the proposed cockpit in a room adjacent to Captain Ziegler's office.

This pushed me to remind him of the makeshift simulator centre I visited in 1976 when I was a student. He laughed and said he remembered well the troubleshooting of the simulator.

This was one of the main topics discussed during the following lunch at the *Le Cantou* restaurant close to Airbus in Blagnac. Located at the end of the runway, Le Cantou was for many years like a second *cantine* for Airbus colleagues and their guests.

Following this visit, the Airbus team came several times to Tunis to explain the FFCC concept. Meanwhile, Garuda signed a full contract for 11 aircraft equipped with Pratt & Whitney engines and keeping the old analogue avionic systems. This move from Garuda solved at least two problems: the A300B4 FFCC was to exist, and there was to be a training centre in Garuda, equipped with the corresponding simulator.

Our evaluation took some time, but finally, we concluded that the most cost-effective solution was to buy the A300B4 FFCC. The number of aircraft was supposed to be three, equipped with General Electric engines.

The approval process and the change of president at Tunis Air delayed the signature of the firm order to a point that the first delivery became scheduled for June 1982, five months after Garuda, but a few months before United (for the B767) and close to 10 months before Lufthansa and Swissair (for the A310).

To ensure full acceptance of the aircraft by the Tunisian pilot union, I organised, in my capacity of Director of Flight Operation (I was promoted to Director of Technical Development and Logistics in April 1980, and from January 1981, Director of Flight Operations), a technical symposium on the recent evolution of the aircraft specifications and operations and the inevitable move towards a generalisation of the two-man crew cockpit.

We asked Airbus to launch the production of the first aircraft as an FFCC, but we asked Airbus to wait for a few months before finalising the flight manuals and the checklists in order to decide if the aircraft would be operated by a crew of two or three. We had several meetings with the pilot union and the flight engineer union; we had meetings with the civil aviation authorities and with the Ministry of Transport.

The same message was coming back: *Provided the two-man crew operation is safe, there is no objection to go ahead, if the unions agree.*

The company could not afford to invest in a new aircraft, which was to last for close to two decades, with a three-man crew while it had already a two-man crew operation running with the B737 and was obliged to start replacing the B727 with modern aircraft with two-man crew (at that time, McDonnell Douglas was promoting its MD80 in the MENA region with a two-man crew). So, we had to go for the two-man crew solution.

I was 30 years old, in charge of the flight and cabin crews and other related flight operation departments; and the future of the Airbus aircraft within Tunis Air was hanging on how I could manage to make the unions accept the two-man crew cockpit, especially as the Tunisian unions were close to the French ones. It was a big challenge, and initially, I found myself with only a handful of pilots supporting the project.

Luckily, the Reagan Committee, in charge of confirming if the two-man crew operation was safe or not, started its work, and during a visit to Denver, Colorado, in the spring of 1981, one of the chief pilot instructors of United Airlines told us that his company would certainly shift to a two-man operation on the B767 because it was very likely that the Reagan Committee would endorse the two-man crew concept.

I was encouraged by that news and came back to Tunis with the intention to push hard and fast to get the authorities' approval. By June 1981, the support base for the two-man crew concept became large, including several flight engineers, and the unions showed an understanding attitude.

Thanks to information received from both Airbus and United, we knew that the presidential committee in the US would give its final opinion by July 1981. Consequently, we decided to organise a symposium and different meetings during the summer period.

After a few weeks, I had reached a formal agreement with the unions for the introduction of the A300B4 FFCC with a two-man crew and the re-training of all the flight engineers to convert them into pilots, according to pre-set planning.

Those who could not succeed in the different tests would be offered jobs on the ground until retirement. A follow-up meeting took place with the Minister of Transport, during which we got the final green light to proceed.

I always remember the president of the pilot union, the late Samir Tabib, saying to the minister:

"We agree with the management. We will help in achieving all the set targets, and we trust that this decision is the best for the company and for the country."

It was an exciting moment, never lived by any other airline in the world. It made Tunis Air and Tunisia true pioneers in advancing the aviation industry. I was proud of this achievement and, at the same time, afraid of the potential hurdles ahead.

Immediately after this meeting, we informed Airbus of our decision, and the whole top management exploded with joy because, from now on, they would have the example of an airline flying an Airbus aircraft with a two-man crew in Europe on a daily basis. Garuda was a good achievement, but their aircraft would be operated domestically in Indonesia and regionally in Southeast Asia, with no visible impact on European airlines and, in particular, Air France.

But the real difficulties only started. First, there was a need to write the training manuals and make them approved by French civil aviation. It was a horrendous task for the poor Airbus instructors and the civil aviation controllers due to the blockages by some people controlled by the unions.

Fortunately, with Bernard Ziegler, we called on Jean Pinet, the head of the training centre, Aeroformation, and former Concorde test pilot who was the first to break the sound barrier. We started a real Kamikaze action with the help of Tunisian civil aviation as a main sponsor and the assistance of one senior French civil aviation pilot, Captain De Castelbajac.

Jean Pinet, with whom I developed very strong relations, did a fantastic job using his reputation and network. Bernard Ziegler used his pushy approach to work inside the pilot community, and I got in touch with the Air France Flight Operations Directorate to ask the Director Captain Massotti, to calm down his management team, who was very vocal against our project.

During our first meeting, Captain Massotti explained to me how Air France was trying to adapt its crews to the evolution of the technology. My conclusion was that Air France had not come out of its rear-guard battle, and the two-man crew would not be there soon. Meanwhile, all the northern African airlines, including Tunis Air, introduced the B737 with a two-man crew, and the wide-body was coming soon to Tunisia.

By the summer of 1981, I had a very unique experience which will never be repeated again under any circumstances, to fly with seven other people, including two Airbus pilots, in an empty A300 FFCC, equipped with only five seats in the cockpit and one sort of stool in the main cabin.

We were lucky, the flight went well, and we managed to rotate through the cockpit to evaluate its ergonomics. We had the opportunity to participate in some critical meetings to freeze the definition of the cockpit, and we were eager to see the end result in real life. These were the pioneering days of Airbus, where a handful of dedicated people could do miracles. Those were the days!

By February 1982, all the obstacles were removed, and the training manuals, as well as the flight, operations, and maintenance manuals, were approved by the French DGAC. The training planning of the flight engineers was ready, and the A300B4 FFCC TS-IMA (MIKE ALPHA) arrived in Tunis on 5 June 1982.

It was supposed to be followed by two others, but the financial situation of the company changed the plans, and Tunis Air remained with one single unit, which was a challenging situation. Due to the innovation in the avionics and the adaptation of several solutions used for the A310, the Tunis Air aircraft was a real debugger for the A310 (Air filter of the avionic bay, instability of ILS, etc.).

During the first six months of operation, we did not stop receiving calls from Lufthansa, Swissair, and KLM about the reliability of the aircraft and the problems encountered.

We had a few difficult moments, but the dedication of the Tunis Air teams and the strong support provided by Airbus, with Bernard Ziegler coming to Tunis on a regular basis, on top of the two months he spent as a line instructor for the pilots, helped us overcome the hurdles. Finally, the aircraft stabilised with a reliability rate close to 100% and became a real workhorse in the fleet of Tunis Air.

As a consequence of the choice of this aircraft, Tunis Air jumped in the front wagon of two-man crew and digital equipment and became the launch customer for the ATEC5000, the first dedicated automatic test bench for digital equipment and member of the IEEE steering committee, in charge of defining the new ATLAS language for ARINC 700 equipment testing, jointly with Aérospatiale from France (as the manufacturer of the ATEC), Boeing, the American DOD, and Lufthansa, plus two known companies manufacturing avionic components.

I was surprised to find our small company among these big names. Unfortunately, we could not pursue our participation because I left the Technical Development Directorate, and no successor was designated to replace me. There was a young engineer who had been helping me with these new high-tech subjects, but I sadly cannot remember what he did after my departure.

One important consequence for Tunis Air, with the arrival of a single aircraft in their fleet, was to decide to do the whole maintenance (except for the GE engines) in-house.

Such a decision for one single aircraft could look silly, but, thanks to special support from Airbus and a clever organisation by the engineering department, Tunis Air succeeded in maintaining the aircraft in-house efficiently. A dispatch reliability rate of 100% was very frequent for long periods and at acceptable economic conditions. In any case, no airline operated a similar type in the region.

The only other solution would have been to send the aircraft to Garuda in Indonesia for maintenance. The mere ferrying of the aircraft, back and forth, would have been very expensive, with a loss of at least three days. This decision to do the maintenance in-house was taken after a visit I did with our COO of Technical and Operations, who was previously the Technical Director, to evaluate all possibilities of cooperation with Garuda.

Besides discovering Jakarta and the Indonesian culture, we established friendly relations with the Garuda management, which allowed us to have frank discussions and conclude that the only possible cooperation was for the training of the cockpit crews, thanks to the flight simulator acquired by Garuda to support its fleet of 11 aircraft.

Consequently, Tunis Air pilots used to travel twice a year to Jakarta to conduct their recurrent and type rating training. I believe they enjoyed that solution for the close to 20 years the aircraft was operated by Tunis Air.

The A300B4 FFCC saga had other implications on our relationship with Air France. Our pilots had been the joke of the block each time they were around Orly (which was daily).

In order to protect Tunis Air pilots, we contacted Air France Flight Operations and a meeting took place, during which I felt some embarrassment within the Air France side.

The only outcome of this meeting was to be patient, and all this would disappear soon. A miracle was in the making at Air France. Surely, after a number of years, a two-man crew solution also became a reality in Air France.

In this respect, it is important to put this union's fight against the two-man crew cockpit into perspective.

The number of persons in the cockpit evolved over time, passing from five in the Boeing 314 Clipper (1938), with a captain, a co-pilot, a flight engineer, a navigator, and a radio operator (this was still the case with the Soviet aircraft up to the '80s), to four in the early versions of the B707 (1957) and the Comet, with the suppression of the position of the radio operator, and finally to three with the more recent versions of the B707 and other aircraft like the Caravelle and the B727, with the suppression of the position of navigator.

I never read about strikes or social movements caused by these successive reductions in the size of the crew.

The situation stabilised at three for some years, but two aircraft, one from the UK (the BAC 111) and the other from the US (the DC9), changed the paradigm by introducing, respectively, in 1963 and 1965, the concept of a two-man crew.

Strangely enough, these two aircraft flew from day one with two men, without any big noise made by the unions. But when Boeing came up with the same concept on the B737 in 1968, the airline community saw some strong opposition, to the point Boeing offered two options (two or three-man crew), with the consequence that some airlines flew the aircraft with a crew of three. God knows what the flight engineer was doing.

But when it came to the wide-body aircraft, the opposition was so hard that Boeing was obliged to build its first 30 B767 with a three-man crew cockpit because it was a wide body, while the B757 was built with a two-man crew cockpit, because it was a narrow body. The two aircraft were sisters, with the same cockpit and with just a small difference in number of seats between the B767-200 and the B757.

It is shocking to see all these intelligent people defending the three-man cockpit for the B767 and accepting it for the B757 as if the risk of killing 200 passengers in a B757 is more acceptable than the risk of killing 220 passengers in a B767. Eventually, Ansett Australia flew with a three-man crew, and United had few aircraft built for a three-man crew.

They were modified after the airline decided to implement the two-man crew following the conclusion of the Reagan Committee. Air France remained one of the two only airlines which continued operating an airliner of that generation—in that instance, the A310—with a three-man crew cockpit specially developed for them by Airbus. Ansett did the same with the B767 for a while until its bankruptcy in 2001.

I personally never understood why this rear-guard opposition took place and why the wide-body aircraft caused more problems than the single-aisle aircraft. For some airlines like Air France and Ansett, the problem was valid for both categories until a global solution was found. I never had a satisfactory answer up to this date. Sometimes, corporatism has reasons that reason ignores.

Coming back to Tunis Air, in a nutshell, the A300B4 FFCC was a driving force which introduced a lot of positive changes within the company.

At a personal level, this adventure helped me learn how to manage a complex project and deal with multiple stakeholders, with conflicting interests and know better the Airbus organisation thanks to Roger Béteille, Bernard Ziegler, and Felix Kracht.

This founding father of Airbus was humble and kind enough to spend time with a young engineer from a comparatively small customer airline and to explain to him details which were not really relevant to the contractual aspects. My relations with Mr Kracht continued for many years.

I also had the pleasure to meet his daughter Barbara, who became the Airbus Media and Press Relations Vice President for many years and who became a close and dear friend of mine. My relationship with the Kracht family extended to Barbara's daughter, Marina, and continues to this date.

During my term at the Tunis Air flight operations, we succeeded to implement some important changes, like making English the working language for the flight crews and to stop the use of manuals adapted from the Air France documentation, to introduce automated flight plans and fuel quantity calculation, computerisation of crew assignment, and, last but not the least, imposing FAA (the US Federal Aviation Administration whose safety rulings are accepted worldwide) rating to all new pilots. This made the airline make a quantum jump in efficiency.

Among the most pleasant activities I had in Tunis Air, I must mention teaching flight instruments at the Borj El Ameri Flight Academy. I had the opportunity to stand again in front of students and transmit some knowledge. The timing was perfect because aviation was in a transition period, between analogical instruments and digital instruments, with the arrival of the Cathode Ray Tubes (CRT) display screen and the laser inertial reference system.

It was challenging but motivating to explain these important changes to students who were acquainted with the classical instruments and who learnt to fly with them. I was lucky to have two promotions of student pilots with big flying experience due to the absence of any hiring opportunity by the main airline of the country. The poor students were waiting for the first opportunity for close to three years, which was a real suffering.

My role, among other things, was to ensure that they were up-to-date and ready to go at any time. Thanks to their flying experience, it was much easier to develop the courses about the new technologies and engage in very interesting debates. One of the students was a lady who became the second female pilot in Tunisia after Alia Menchari, who was the first in the country and among the very few first in Africa and MENA region, with Taghreed Akacha from Jordan.

I was shocked by the situation of these future pilots—to be on a waiting list for a few years without a clear visibility. I thought it was a crime. I suggested to finish their training and putting them on the international market to give them an opportunity.

Unfortunately, it was necessary to have a European or a US licence to be accepted by international airlines in foreign countries. This was the main reason which pushed the management team of flight operations to introduce the complementary training at Sierra Academy of Aeronautics in the US, with the B737 type rating done at the united training centre in Denver, Colorado.

This move made the Tunisian pilots marketable, leading to today's situation, where there are more Tunisian pilots flying abroad than pilots flying with Tunisian airlines.

One of the key factors in the build-up of this stupid situation was the lack of confidence of some people within the airline and in the aviation sector in the fleet plan and growth of Tunis Air. It was forecasting the doubling of its fleet size by the end of the century and the need for a lot of pilots. The sceptics succeeded to run the show for a while and consequently to close the school.

This led to a wide-open road for alternative flight training, which served those who had the financial resources to pay for their children. Some Captains succeeded in training their sons or daughters and have them ultimately recruited by the company. These captains were among those who were promoting corporatism within the company by defending their interests without any consideration for the interests of the company and their countrymen.

Luckily, the government reacted quickly and put the academy under military control.

4.4 Tunis Air at the Frontline (Front Line-Stevie Wonder)

One of these major events, which quite often make history, took place in September 1982, and I had the privilege to be a prominent witness: the transfer of Yasser Arafat from Athens to Tunis after his evacuation by the French forces from Lebanon, following the ceasefire agreement obtained by the US envoy Philip Habib between Israel, which invaded Lebanon in June, and the PLO and other Palestinian organisations.

I was head of flight operations, and as such, on 2 September 1982, at 10 am, I was informed that I had to prepare a B727 to be dispatched to Athens and bring back a VVIP. This was the only information I was given at the time. A few minutes later, the president of the airline asked me to come with him to an important meeting with some ministers.

Twenty minutes later, we were in a large meeting room, at the Ministry of Defence. Three ministers were present: Defence, Interior, and Foreign Affairs, as well as the head of military intelligence, the head of a similar function at the Ministry of Interior, and the chief of the Air Force. The president and I felt our presence awkward in this huge room.

The Minister of Defence started by saying that, of course, all the persons present must respect a strict confidentiality obligation on the whole content of the discussion. Then, he explained that the president of the Republic of Tunisia had proposed to give asylum to all the Palestinian fighters leaving Lebanon, after Philip Habib had brokered a ceasefire between the Palestinians and the Israelis, as well as to the top officials of the PLO, including the leader, Yasser Arafat.

Now, it is the turn of Mr Arafat to be brought to Tunis. A very complex operation already took place to get him out of Beirut, and now it was our duty to transport him from Athens to Tunis in one of our Tunis Air aircraft. There was still a risk that the Israelis would try to kill him. Therefore, all that would happen from then on had to be done under strict security measures.

All the flight details were to be passed to the Greek and Italian civil aviation authorities through diplomatic channels, and an officer of the Air Force would be present in the Tunis Air Operation Centre to ensure proper coordination with other departments.

A brief but detailed discussion followed. When my turn came, I commented, naively, that I had to designate a cockpit crew and a cabin crew. Therefore, what could I tell them? The answer was to tell the truth, but we needed to be sure that they would keep the information confidential, up to the end of the mission.

This was the trick: call somebody, ask him or her if he/she accepts a mission during which he/she could be shot down by the Israelis, and tell him/her to say nothing to his/her family until his/her return, if he/she returns. Noticing my concern, the Minister of Foreign Affairs gave a reassuring statement by saying that the Tunisian government is in close contact with the US, French, Italian, and Greek governments.

The whole plan had been endorsed by these countries, and the Israelis had promised the US government not to derail the process by any unilateral action. Therefore, the risk was minimal.

On our way back to the airline headquarters, I proposed to the president of Tunis Air to let me find a solution within an hour. As soon as I arrived at my office, which was next to the airport, I called the chief pilot, Captain Mahmoud Ben Youssef (RIP), who was a close friend and a very straightforward man with solid guts. I told him the full story.

His answer was:

"I have already selected the full crew. I'll be the captain, and the other people will be here within an hour."

At around 2 pm, I had nine persons in my office: one captain, one co-pilot, one flight engineer, one ground engineer (whom I called personally), one purser, and four flight attendants. They were all determined and ready to go at any time, and fully aware of the high confidentiality required. I called the president to tell him that the crew was ready to go. I just needed the okay to give the go-ahead to the crew.

The take-off took place the next morning. The only thing I remember is my meeting with Captain Ben Youssef in the briefing room and the long hug.

Years later, his wife told me that, on that particular day, she did not know anything and was thinking it was a routine flight. But she was surprised by two strange comments her husband made before leaving the house for the airport:

"Take care of the children and promise me not to marry another man if I die."

From that moment on, I switched to the simple role of a monitor, through the different available tools and through the information received from the military and interior authorities.

I spent close to one day and a half under continuous stress, which was further complicated by the information received during the flight between Athens and Tunis that Greek and Italian jet fighters were escorting the aircraft until a few miles from Tunis Carthage airport, to avoid any attempt by Israel to shoot down the aircraft. During all this time, I was thinking of what I could say to the families of the crew members if that horrible scenario were to become true.

I had a twinge each time I visualised the face of Aroussia, the wife of Captain Mahmoud, who is like a sister to me, standing in front of me while I was announcing the horrible news to her. I spent the whole time, between the take-off from Tunis and the landing back from Athens, shuttling between my office, the Operations Centre, the president's office, and the Flight Control Centre of the Civil Aviation.

Due to the need for on-time information collection, some people holding key positions in running the daily operations were put in the loop, and I had the possibility to exchange with them with only a few limited words.

At the arrival of the flight from Athens, in the afternoon and during taxiing in, we were a small group waiting at the VIP lounge. But for some reason, I still fail to understand, up to today, the loud noise of applause which started covering the tarmac.

With the head of the security forces at the airport, we got out of the lounge on the tarmac side, and we were surprised to see hundreds of people from different entities working at the airport, approaching the aircraft, which was still taxiing, escorted by four police cars. We could not understand how the news had spread so quickly.

The aircraft was immediately surrounded by security forces, which put the crowd at bay, while two helicopters of the army were circling above our heads and three fighter jets from the Tunisian Air Force flying in the vicinity of the airport. It took some time to prepare all the logistics around the aircraft before allowing Arafat to disembark and get a warm welcome from the Tunisian officials and members of the Palestinian leadership who were already installed in Tunisia.

I was so exhausted physically and psychologically that, after welcoming the crew and thanking each member for their contribution, I went home, escaping from all the PR exercises and the official debriefings. I just called the president to tell him I was going home. The crew was received later by President Bourguiba and received a thank-you letter from Arafat.

The crew was the real hero of this special operation because, despite the commitment to the US, Israel had envisaged several possibilities to kill Arafat, knowing that there would be no serious consequences for them. An attack against the aircraft was one option.

Finally, the serious attempt to kill Arafat took place in October 1985, when Israeli F15 flew all the way from Israel to Tunisia and, in particular, to Hammam El Chot, south of Tunis, where they bombarded the Arafat headquarters and killed dozens of Palestinians and Tunisians.

Once again, death came from the sky. It was the first time that the US abstained in the Security Council during a vote against Israel. But everybody in Tunisia was convinced that the US were either part of the attack or at least aware of it. History will judge.

4.5 Developing and Reforming the Airline (Bohemian Rhapsody-Queen)

In early 1982, in addition to flight operations, it was decided to put me also in charge of all ground operations, including all out-stations. These new responsibilities allowed me to get a full view of the airline's daily operations (maintenance, passenger handling, cargo handling, flight operations, fuel management, and catering).

Taking the opportunity of the existence of ambitious plans to modernise many aspects of passenger handling, I decided to speed up the implementation of all the existing projects, and in particular, the automation of passengers' check-in, through the system called RayCheck, developed by Raytheon, as well as the automation of the weight and balance computation, plus other functions.

Thanks to the enthusiasm of the teams in charge of these different projects, we succeeded in implementing the new systems within a relatively short period of time.

I was adamant about changing the way Tunis Air was working, and the images of the manual processes I saw during my first flight in 1970 were still haunting me. By early 1983, many of the actions related to the daily operations of the airline were automated. It changed the workload and the lives of hundreds of employees. My great satisfaction was the positive reaction of the personnel at Tunis Carthage airport and in the stations.

During my first seven years at Tunis Air, I encountered two contradicting aspects among the personnel. The first one was a strong corporatism, particularly amongst the flight crews, and the second one was a friendly and enthusiastic attitude, with a lot of dedication among these very same personnel.

As long as the two attitudes were balancing each other, the company was doing comparatively well and succeeded to overcome many difficulties. But when the corporatism became the dominant factor, the airline started collapsing. This was, and is still, true for most of the airlines.

One of the main prominent factors I noticed in all airlines around the world was the strong corporatist spirit of the pilots in Europe, and mainly in France. Each pilot considers himself as the heir of St Exupéry.

It takes a full understanding of the global balance of forces within an airline and a very diplomatic and, at the same time, a straightforward approach to explain to the pilots their real role in an airline. Meanwhile, the test pilots working for the aircraft manufacturers show a great sense of humility and are fully integrated with the rest of the employees.

From the end of 1982, however, I started facing active opposition from a group of pilots who were against most of the reforms (two-man crew, English as a working language, FAA type rating, automation of crew assignment) and were challenging all decisions related to the refusal of releasing some pilots to fly as co-pilots or captains due to their inability to reach the required level.

The motto was to *give them more training.*

This was one of the strongest expressions of stupid corporatism (I found that this expression was often used in many airlines). We had the first critical case with the type rating on the A300B4, where I was requested to give more simulator sessions to a captain who could not make it. Despite the additional training, the captain did not get his type rating.

I was lucky to have a good management team with a strong chief pilot. We managed to implement all the reforms and stood firm about the no-release of certain pilots. Unfortunately, some of the instructors were influenced and manipulated by the opponent group, and the top management of the airline was not standing by our side.

When the pilots called a strike to demand the departure of the chief pilot, the president of the airline accepted their demand and asked me to fire him. I refused and submitted my resignation. He was happy with this outcome. He appointed me director in charge of strategy, development, and information systems, separated flight and ground operations, and appointed a captain as flight operations director.

This captain and his new management granted further training to the pilots who were not initially released and eventually released most of them. Of course, all the regulatory aspects were respected, but how much did it cost the company in terms of simulator and flight hours?

Unfortunately, this move sent a wrong message to the whole company, and I made a point to send the president a letter stating my position vis-à-vis all this saga and warning him of the consequences. A few months later, one event confirmed my warning, and the president called me to ask me not to escalate the matter and promised me to undertake all the necessary actions to correct the situation.

Unfortunately, this was the accelerating element of the sequence of problems the airline was to face in the future. The different unions took control of the management of the airline. This happened in many other airlines worldwide with very well-known dire consequences.

Despite this disappointing episode, I decided to do my best to serve the company in my new position. The circumstances played in my favour. On 2 January 1984, a new president was appointed, in the middle of an escalating economic crisis hitting hardly the country and which reached its peak with the *Bread Riots* which took place between 27 December 1983 and 6 January 1984.

These riots were triggered by the sharp increase in the price of all grain-based products and, in particular, bread, as per the stabilisation programme agreed with the IMF. They swept all the regions of Tunisia and caused the death of 143 persons, in addition to more than a thousand injured. In order to stop the riots, the president of the republic had to declare, on TV, that all the increases were cancelled.

These events had dramatic effects on the political landscape in Tunisia for decades to come. The social transformation lived by Tunisia was not accompanied by political transformation, creating a serious discrepancy and generating great disappointment.

The newly appointed president of the airline was one of the long-serving executives; his former position was COO of Commercial, Financial, and Administration of the company, and he was fully aware of the situation of the airline. In particular, he had a clear idea of the difficult financial situation and the need for an effective and rapid recovery plan.

For historical reasons (competition with his colleague COO technical and operations, and some regional preferences), he did not like many directors and senior managers originating from the technical directorate, including myself.

As a strong sign of this hate, he decided, as of 4 January 1984, to cut the salaries of the other COO, of another director, and my salary, by a sizeable chunk representing all the technical allowances. On that date, I understood that my time had come to leave for other shores.

Nevertheless, I must admit that besides this unjustified action, he showed me respect and gave me his full confidence to undertake a serious reform of the airline and elaborate a recovery plan.

During the first six months of 1984, a complete restructuring of the company took place, coupled with a recovery plan, covering 10 years, reducing the size of the fleet by three units, balancing the activity in favour of scheduled flights, improving the quality of service, signing wage agreements with the four unions (pilots, cabin crew, maintenance technicians, and the rest of the personnel), stipulating linkage of the salaries to improvements of the productivity (better flight dispatch reliability, less fuel consumption, increase of commercial revenues, higher customer satisfaction, etc.).

I was the coordinator for all these actions, dozens of people were involved and, in particular, my two pillars or *flankers* as it is called in rugby: Hamadi Thamri, a graduate from ENSEEIHT engineering school in Toulouse, for the technical aspects, and Moncef Ben Dhahbi, a graduate from the Faculty of Economics of the University of Tunis, for the economic aspects.

All the targets were achieved in a record time, the new organisation was implemented and fully running, the recovery plan implemented, and the airline was able to rebuild its cash reserves, reduce the cost of different departments, and improve the efficiency of its operations.

Among the results of these actions, we can mention the sale of two B727s out of three planned to Air 1 (nothing to do with the Italian Air One), a US-based airline which tried to launch an all-first-class service out of St Louis. The last aircraft, for which a deposit has been paid, could not be delivered nor a stock of some spares, due to a sudden and surprising shortage of cash at the buyer.

Despite this last-minute problem, the transaction was profitable for Tunis Air, which was able to generate the necessary cash to optimise its fleet. Some cost-cutting actions generated more cash savings than expected and compensated for a good portion of the missing part of the price of the third aircraft.

We can also mention the signing of the first wage conventions in the country, linking salaries to productivity and the launch, full gear, of the automation of all the operations of the company, which were important cost savings generators.

During the negotiations with Air 1, I came to know that one of their advisors (or board member) was Eugene Cernan, the US astronaut who was the 11th walker on the moon. It was a surprise to find this celebrity in an airline environment. Following an exchange of souvenirs with the representative of Air

1, who was present in Tunis to finalise the negotiations, I gave him an extra present for Mr Cernan to make him remember our small country.

Nearly two weeks later, and to my great surprise, the representative of Air 1, who was back in Tunis, brought me a beautiful photo of an astronaut walking on the moon, with the following words:

"To Karim, with best wishes from Eugene Cernan," with the date of December 72.

This photo is still displayed at the dental practice of my son in Dubai. I then understood why I had been asked, some time ago, about my son's name.

On 2 March 1984, Airbus launched the A320 with technological breakthroughs which were real game-changers in the field of civil aviation (fly-by-wire inspired by Concorde and military aircraft, side stick or joystick in lieu of the normal control yoke inspired by video games, new flight envelope protections, and many more). Just a few days earlier, Boeing had flown its new version of the well-established B737, the dash-300.

These two events made McDonnell Douglas (MDC) very nervous. As a result, they decided to strengthen their sales campaign for the MD80 with Tunis Air, and we had two years of close exchanges with McDonnell Douglas, which I already knew well. This very aggressive sales campaign by MDC followed the demonstration tour they had organised in the spring of 1983, covering many countries in North Africa and the Middle East.

The MD80 came to Tunis, in May 1983, under the captainship of Charles (Pete) Conrad, the former astronaut, who was the third man to walk on the moon on 14 November 1969. During the two days the MD80 was in Tunisia, we had two demo flights and several meetings, and of course, lunches and dinners. I was often seated next to Pete Conrad, and we could discuss several subjects. I refrained from asking questions about his journey to the moon.

But other colleagues did, and he was ready to answer. Honestly speaking, I did not hear anything new, but the way he expressed his feelings and detailed his reactions when he was up there made the moon adventure look more human.

The most important details which attracted my attention were those related to the specifications of the spacesuit and the multiple functionalities it had to accommodate human needs, protect the astronauts, and serve as a holder for a sort of tool kit enabling the performance of certain technical actions. Before one dinner, we were walking on the beachside of a famous hotel next to Carthage.

We had an interesting exchange about the history of Carthage and the famous general Hannibal. I was impressed by his knowledge about some important facts and events related to Carthage, and he made a comparison between Hannibal and Alexander the Great.

In particular, I remember him saying:

"The world of today, as a civilisation and a political system, is a consequence of the defeat of Hannibal against Scipio the African, followed by the destruction of Carthage, much more than of the victories of Alexander against all the countries he conquered."

"Some cities and regions still bear his name, but the civilisation and the lifestyle we have today are a direct inheritance of the Roman Empire, which survived under different forms until the beginning of the 19th century, which saw the end of the Holy German Roman Empire."

I was bluffed by his knowledge of history and his acute analysis. We also had an interesting exchange about music and about some famous rock and roll bands from both the US and the UK. We agreed that the music of the late 60s was largely influenced by the civil rights movement, the youth, and the anti-Vietnam War protests. It was a pleasure to recite, together, certain lyrics from famous songs.

Here again, I discovered a man who was really down to earth, even if he seemed to have a certain ego. But our best discussion took place in California, a few months later when, with members of my team, we paid a visit to Long Beach, in order to finalise the evaluation of the MD80.

Following a meeting, Pete Conrad invited us for lunch in a Mexican fish restaurant. When we arrived in front of the restaurant, we saw a fantastic Harley Davidson.

I expressed my admiration for the beauty of the machine, and he answered with a laconic sentence like:

"Yes, very beautiful, but it can be dangerous in the hands of an inattentive rider."

We continued our discussion about motorcycles. We spoke about the movie *Easy Rider*, Steve McQueen and the stunt with the motorcycle, and about the Hells Angels and their cult of motorcycles. He was very knowledgeable of mechanical sports and did not hesitate to explain to me a few details. The food was fantastic, and crab meat was available in huge quantities and in different forms and shapes, to the point, it is still remembered by my Tunisian colleagues.

When we left the restaurant, another Harley Davidson motorcycle was there, but this one was a *vintage* of the 50s, according to Pete Conrad, who was a real connoisseur.

He confirmed his admiration for this machine by saying:

"I love this one."

15 years later, Pete Conrad died following a motorcycle accident. His death affected me tremendously.

I keep a vivid memory of these three full days I spent with this remarkable man. He was an intelligent, courageous, straightforward, and easy person who had a huge culture and a great sense of humour.

I still had another astronaut coming my way, but it took close to 26 years before I met him.

One of the most significant discoveries I made, thanks to my multiple contacts with aircraft and engine manufacturers, was that there is a small number of Middle Eastern and North African employees working for these companies, many of them being Egyptian.

Boeing, GE, P&W, and Mc Donnell Douglas each had a few MENA citizens among their personnel, but it seemed that Airbus had none; later on, I discovered that they had a Lebanese salesman, but he did not stay at Airbus for long.

I was curious to know the size of the Middle Eastern and North African community working for the main manufacturers and tried, through different channels, to get an answer, but it was difficult until I came across a Jordanian journalist, in 1984, who was writing an article about the Arab diaspora of aerospace employees at managerial positions, who told me that the total number did not exceed 20.

I cannot confirm this figure, but it seemed very plausible. From the people I knew, personally, I concluded that the majority were in sales and marketing. It took the industry some time to become more open to Middle East and North African citizens, and it also took the concerned countries some time to educate and train enough competencies to attract the attention of the aerospace industry.

Nowadays, the number is much higher. In addition, many countries in the region have developed local aerospace ecosystems which employ thousands of people. Added to the globalisation of the major manufacturers and suppliers, citizens of the MENA region now make up a respectable part of the global aerospace workforce. If we add the airline ecosystem, the MENA region becomes a real powerhouse. I will develop this aspect later.

4.6 Managing a Stupid Survival Crisis (Under Pressure — Queen)

As from the second half of 1984, Tunis Air started improving its cash situation and its operations. Unfortunately, the political situation of the country started degrading and generating internal problems within Tunis Air, due to the connection of some prominent political figures to some top managers of the company, in particular the strong relation between the wife of President Bourguiba and the president of the company.

Luckily, the train of reforms had continued to advance and was on track, but the authorities started slowing down the president and questioning several of his decisions. By November 1985, the president was fired, put in prison, and replaced by somebody with an aviation background, but unable to run a complex corporation like Tunis Air.

He was under the full control of the minister and was instructed to apply the political agenda of the entourage of the President of the Republic. Between November 1985 and May 1986, the situation of the company degraded so much that the new President was himself replaced by June 1986.

During this turbulent period, I was removed from my position as director, along with 11 other colleagues, for no apparent reason.

I asked the President to give me the reasons for my dismissal, but his only response was:

"Wait for better days."

This was the stupidest answer I ever heard. I was then appointed chief engineer at the technical directorate.

During the weeks following this dismissal, many of my colleagues were summoned by the financial brigade of the national police, which was investigating potential embezzlement within the company. The whole story was to justify the firing and imprisonment of the President. It generated injustice and drama, which reverberated for many years.

I was never called for any questioning, but I decided that the time had come to leave Tunis Air and go to work under other skies. I was appalled by the short-sighted behaviour of the authorities, which favoured taking revenge and destroying the company in lieu of supporting it in its efforts to restructure, improve efficiency, and generate cash.

1985 was an *annus horribilis* for the country, the company, and also for me. Even before the troubles of Tunis Air, which affected me directly, I lost my father in August 1985, following a heart failure, at the age of 71. To get out of the negative spiral and reflect properly, I took, by the end of January 1986, four weeks off, to travel to the Gulf and to Europe.

I went to Dubai and Abu Dhabi, to see the evolution of these two cities I had last visited in March 1980. I came back to Tunis through Bahrain. Before boarding the Gulf Air flight to Tunis, I met a prominent figure of Bahrein, Mr Youssef Al Shirawi, who was the Minister of Development of Bahrain and the father of industrialisation of the country. He was inspecting some projects at the airport.

He was introduced to me by one colleague of Gulf Air. We spent a few minutes chatting about Arab Aviation, and he explained to me that he was closely following Gulf Air, but he had a serious concern, which was the need to accelerate the nationalisation of the workforce of the airline, in all the departments.

He asked me how Tunis Air succeeded to achieve it and concluded by adding:

"I'll be happy to see a lot of Arab executives working for the big aircraft and engine manufacturers."

These comments rang a bell and pushed me to contact Airbus, Boeing, and others, to check if they had job opportunities for me. I then travelled to Europe to meet the management of some companies, including Airbus.

I had some attractive propositions, but Airbus went further by offering me the position of regional sales director, confirmed by one contract signed unilaterally by the company, detailing all the conditions and valid for three months. I came back to Tunis with the clear decision to leave Tunis Air and to go for an expatriation.

Consequently, I resigned in March 1986 to join Airbus in Toulouse, from May 1986, as a sales director for Algeria, Egypt, Libya, and Morocco. Tunisia was excluded, contractually, to avoid any conflict of interest.

My choice for Airbus resulted from a number of discussions I'd had with many aerospace companies and was facilitated by the courageous attitude of Airbus, which gave me a signed contract, and also by the fact that my wife and I knew very well the city of Toulouse, and that my son could easily be educated in France.

But there is one person who played an essential role in convincing me to join Airbus: it was Jean Pierson, the new CEO of Airbus, who took over from Roger Béteille in April 1985.

Jean Pierson was a French citizen born and raised in Bizerte, Tunisia. He lived in the country till the age of 23. I met him for the first time in 1981 when he was the Head of the Aérospatiale factory in Toulouse.

At the time, we were in the process of defining the initial provisioning for our A300B4 FFCC. Roger Béteille had suggested that Aérospatiale Toulouse could help Tunis Air with some AOG (Aircraft on Ground) situations, where it could need a quick supply of parts that were unavailable in Tunis Air inventory. J. Pierson was very helpful, and our Tunisian link created a nice atmosphere.

We became friends, and he was invited with his wife to be on board the aircraft during the delivery. He spent three memorable days in Tunisia (it was the first time he returned to the country after he left Bizerte in 1963). As the new company CEO, he encouraged me to join Airbus and promised to facilitate my installation in Toulouse.

His attitude removed any hesitation to leave my country and start a new life, for me and my family, in another country. For the next 12 years, Jean Pierson played an important role in my career and in my life.

I started my new job, with Airbus on 12 May 1986 and opened a new and exciting page.

I joined Airbus as a Tunisian citizen, and I had to go through a complex immigration clearance process.

Thanks to the help of Airbus, I did not waste a lot of time, but I had a close encounter with the multiple hurdles through which other immigrants, and particularly those coming from North Africa, have to go through.

While I was getting my residence visa and my work permit, I remembered my interview with a colonel from the French army, in January 1972, at *Caserne Cafarelli* in Toulouse, in the frame of my application to pass the examination to enter Ecole Polytechnique of Paris, which is part of the French Ministry of Defence.

The question of the officer was:

"Would you like to pass the exam as a French citizen or as a foreigner?"

It very much surprised me, but he explained that due to the services rendered by my father to France, during his career in the French army, I was allowed to request French citizenship. After consulting with my father, I chose to pass the exam as a foreigner.

I never regretted this choice, but after close to five years at Airbus and despite the big help from my management and the concerned departments, I started facing a lot of difficulties to travel, due to the multiplication of the number of countries imposing visas on Tunisian nationals. This visa requirement was valid in most of the countries in the Middle East region, in addition to the European countries and the US.

After a serious delay in getting a visa from one country in the Gulf where I was supposed to travel with Jean Pierson, he called me in his office and suggested me to request a French passport.

He underlined that he accepted me, from day one, as a Tunisian citizen, and he was not pushing to become French, but he feared that I could be seriously disturbed in my job by the continuous requests for a visa, especially since the Tunisian authorities refuse to provide a second passport. He was right in his appreciation of the situation.

Many times, I had to cancel or delay a trip, because of the visa. In addition, on many occasions, I had to decide on some trips within 24 hours but was not able to travel due to the lack of time to get a visa.

20 years after my refusal to become French, I got French nationality. I got two French passports, and with the Tunisian one, I had all the necessary flexibility to get visas on time. But I hate to consider the French passport as a simple commodity.

I have a lot of connections with France, from my great-grandfather, who fought side by side with the French army in Crimea as part of the Tunisian expeditionary corps, to my father and my uncle, who were in the French army, ending with me and my contribution to the French and European economies.

My relationship with France encompasses a large spectrum of aspects, from language to culture, to the knowledge of the history and the geography of the country, up to politics and gastronomy. I am not an alien vis-à-vis France, but I am not *truly French* because I kept my Tunisian name with its Arabic and Muslim roots, and also because all my roots and my heritage are in Tunisia.

I can interact with the French people like a true Frenchman, but I don't respond positively to the criteria of being part of the Judeo-Christian civilisation, which is supposed to be the basis in France. This concept of Judeo-Christian civilisation is questionable because it ignores a lot of other contributing civilisations and religions (the concept of the Mediterranean could be more convenient) and deserves a serious debate.

I have no intention of reneging on my origins and my real Tunisian identity, nor do I have any intention of not assuming the French part of me.

After just three weeks of familiarisation with the different departments of the commercial directorate and the basic processes, I started travelling to Algiers and Casablanca. I had never visited Algiers, and I had never been in touch with the Air Algérie management, but I have been to Casablanca several times, and I knew all the management of Royal Air Maroc. Consequently, my sales campaigns seemed to be easier in Morocco than in Algeria.

The reality proved the opposite, and we started negotiating a Memorandum of Understanding for three A300-600s, with Air Algérie, by the second quarter of 1987, while in Morocco, it was not possible to consider any change to the Boeing fleet. It is true that the Boeing VP sales for North Africa and the Middle East were Moroccan (Seddiq Belyamani).

I knew him from my time at Tunis Air when I used to go to Seattle. We had a friendly relationship, full of mutual respect, which we kept until our respective retirements. Very quickly, I started focusing my attention on Algeria and Egypt, but without neglecting Morocco. We must never give up!

While I was trying to conclude sales, I wrote a long and comprehensive report to Jean Pierson, the SVP of Commercial Stuart Iddles, and the VP of Sales, Ranjit Jayaratnam, in which I described my perception of Airbus as a customer and my new perception as an employee.

This report generated internal debates and led to some friction with some people, particularly in the marketing department. Jean Pierson and Stuart Iddles endorsed my conclusions and requested some changes, both at the structural and functional levels.

I was proud of this outcome, which was decided during a dinner around a big couscous (a North African dish very much favoured by Jean Pierson) at a famous Algerian restaurant in Toulouse. However, to strengthen my credibility, I had to achieve sales. This was my target for 1987.

I was in the middle of my different sales campaigns in 1988 when a bombshell exploded on my way and was close to shattering my life. In Tunisia, a court condemned me to one year of jail and a one-million Tunisian dinar fine for complicity of corruption linked to the operation of the B727 aircraft sale. It was a total shock. I have never been summoned to any investigation, and nobody from Tunis Air ever called to warn me.

I was in contact with the new airline president, appointed in June 1986, and he never mentioned anything about any investigation. Above all, the sale took place in full compliance with all the rules and regulations (call for tender, comparison of the offers, aircraft sold at a price higher than the purchase price, etc.), and was approved at different levels, from the executive committee to the board of directors, to the ministerial committee, and the prime minister's controlling body.

Luckily, I had copies of all the relevant documents. Before leaving, I was suspicious of possible dirty tricks and decided to copy all the important documents, attest them with the police office next to the company headquarters, and keep them in a safe at a confidential location known only by me, my wife, and my friend Mohamed Cherif. That was my guarantee against such a baseless accusation.

Following a quick investigation, it appeared that a lawyer appointed by the minister to defend the interest of the company, against any party which could have caused prejudice to Tunis Air, found himself without solid arguments to sustain the accusations against the former president of the airline, who was fired in November 1985. Consequently, following some advice from one of the directors, he decided to use the sale of the B727s as a key for his accusation. How?

By claiming that the procedures were not respected. By a miracle, all the relevant documents related to the sale of the B727 had disappeared from the archives of my previous Directorate, even though all the documents, which were sorted by date and subject, were put in there a few days after the delivery of the last aircraft to the buyer, in my presence and the presence of at least three other people, who confirmed their disappearance.

Some copies and all the documents related to the sale of the spares remained in the archive of the technical directorate.

In a short period of time, without questioning the more than a dozen people who contributed to the transaction and by showing only a limited number of documents, a judge decided the fate of innocent people. This was the result of the deliquescence of the political landscape in Tunisia in the mid-80s.

I tried my best during my more than nine years at Tunis Air not to be involved in any political game, focusing my attention on how to make Tunis Air a modern, efficient, and profitable company.

I believed that it was my duty to serve my country without taking into consideration all the visible shortcomings at all levels, which were harbingers of disasters to come. This was my biggest mistake: believing that I could work out of any political influence in a third-world country. Later on, I discovered that this was also true in developed countries, but with more subtlety.

Thanks to all the elements I had in my possession, I engaged a procedure against the accusation and succeeded very quickly to have the judgement cancelled and all my rights restored.

During the period of time this fight lasted, I had unwavering support from my family and, in particular, my wife, as well as from Jean Pierson and all Airbus management, my friends, and particularly Mohamed Cherif, my study companion since the Fermat days and my close friend and colleague, and some of my ex-Tunis Air colleagues, in particular, Hamadi Thamri, whom I dragged to join me in Tunis Air and who also became a loyal friend.

Jean Pierson even called the president of Tunis Air, who explained to him the situation in a very transparent manner, minimising any serious risk for me. Unfortunately, he was not right.

This ordeal allowed me to know my true friends and the hypocrites, as well as to better understand the danger of the political game and how one can easily become collateral damage without knowing it.

This crisis, which was a serious threat to my life and my career, taught me several lessons and made me become very strict in respecting processes and handling key documents. This helped me, years later, to avoid similar situations.

In addition, this unfortunate event made me allergic to any injustice and any betrayal. This could explain some of my actions and positions during the second part of my career.

The only sad side of this unjustified saga was my inability to attend the funerals of my stepmother, my sister-in-law, and my nephew, who died in a car accident in 1988, due to the advice of my lawyer. He recommended I do not travel to Tunisia and wait till all the necessary steps for a review of the judgement were completed.

It was a double punishment, and I kept a very strong sour taste about all this.

On a more positive note, Tunis Air gave me the opportunity to meet and work with fantastic people, who were crazy lovers of aviation and who were dedicated to their job beyond imagination. Many of them became true loyal friends to this date.

5. Practicing the Aviation Job Worldwide (On the Road Again-Canned Heat)

Despite the strong support of Jean Pierson and my management, my arrival at Airbus provoked contrasting reactions. I was supposed to be the first *Arab* to be recruited (I quickly learnt that there was another one before me, but he left the company relatively quickly and did not leave good memories). Since most of the employees were French, and some of them were repatriated from Algeria, there were still strong feelings about North Africans and Arabs.

Therefore, I had to be very diplomatic and avoid any contentious political discussion about the Algeria war. But I was much more shocked by the very negative reactions of the management of the marketing department, which was British. Their comments on my report on comparative perceptions of Airbus from inside and from outside were very harsh and, to some extent, insulting.

Luckily, they were not shared by the Senior Vice President of Commercial, who was also British, and, above all, they were rejected by Jean Pierson, who approved my report. It has to be underlined that at the same time, the French and the Germans accepted the content of my report.

Then, there is a valid question:

"Why did the Brits react so strongly?"

It took me time to find the answer.

The reason was that the British were controlling the commercial directorate and, in particular, the marketing department. Since my report recommended a number of major changes to the organisation and the processes in Commercial, the British felt it was a threat to their leadership.

After a few months, I became a good friend of the head of the marketing department, Adam Brown, and he became a strong supporter of the proposed modifications. Our convergence of points of view was greatly facilitated by our common passion for running.

I started my career at Airbus with a lot of enthusiasm and a lot of hopes. In order to simplify the understanding of the succession of the events, I divided my professional life in Airbus into three main periods: The G.I.E., the Corporation, and the Group.

5.1 Airbus GIE (A Group of Pioneers — 1986–1998) (Simply the Best-T.Turner)

I can say I was really lucky to join Airbus when it was a G.I.E. because it had all the advantages of a dynamic company without all the red tape of big corporations, while using the capabilities of the greatest corporations in Europe.

This structure allowed Airbus to control all the major components of the business with a relatively small group of people who knew each other across all the divisions and had a strong feeling of solidarity and motivation to ensure the success of the manufacturer.

Airbus was able to decide the type of product it wished to produce and sell, coordinate its design and production, define its commercial conditions, sell it, and support it. The famous four partners were responsible for their share of the work and were accountable for everything vis-à-vis Airbus. It was true that the profits were given to the partners, with Airbus keeping enough money to cover the company's operating cost plus some little margin.

This scheme had many advantages, which allowed Airbus to do in 30 years what Boeing did in more than 100 years. There was a special spirit in Airbus, which I called *the Pioneering Spirit*.

This spirit can be summarised by President Obama's slogan: *Yes, we can.*

Nothing was impossible for the Airbus people. Even if there was zero chance to sell, the whole team travelled to try to sell. From the salesman to the customer engineer, to the test pilot, up to the CEO, everybody contributed to the sales campaign.

We alternated our presence within the airline to the point that we gave the impression of a constant presence in the field: Airbus was always present. There was no small customer to approach and no small number of aircraft to sell.

People were not counting weekends and holidays spent on the road and were not paid big salaries compared to the competition and even to the partners. There were many surprises when we compared, from time to time, with our colleagues from the partners. It seemed that the name itself and the challenges were enough motivators to lure the candidates to join this company. I was one of them (I refused a higher salary with another company to join Airbus).

This attitude was very common and contributed to creating this unique ambience and unique spirit, which allowed Airbus to do miracles. I was personally further convinced about this spirit when I discovered the salary of Jean Pierson himself. Following a discussion within my team about the bonuses following a very successful year, I went to see Jean Pierson to ask him for some salary increases.

He listened to me and then got his pay slip from a drawer in his desk and gave it to me, asking me to read it carefully. I was shocked by his reaction, and I quickly read the paper.

Without asking me any questions, he commented:

"Did you see the difference between my salary and the highest salary in your team? Is it too big?"

I must admit, I was surprised by the level of the difference (which was not so huge). Despite his surprising attitude, he increased the bonuses of some members of the team.

Therefore, within Airbus G.I.E., we always worked as a team: a reduced team comprising the sole persons in charge of the customer (salesman, marketing manager, contracts manager, customer engineer), or a medium team enlarged to some engineers, customer support managers, and test pilots, and finally, the extended team, which could include senior executives and even the CEO.

The structure of the G.I.E. allowed Airbus to have a certain immunity to political influences, leaving this burden to the partners, each one with the government of his country. This did not stop Airbus from asking for some support from one or the other of the partner countries.

But the company rarely suffered direct interferences from any of the governments. This was a beautiful setup, at least in theory, having the clout of four countries, among the richest in the western hemisphere, without suffering from the red tapes of the political authorities.

In reality, this was valid for a relatively long period of time (from the early '70s, up to the departure of Jean Pierson in March 1998). A combination of newness, pioneering spirit, strong will from Europe to break the US monopoly in civil aviation, and, above all, to fight for the survival of the aviation industry in Europe made this situation possible and gave us an opportunity to work in an ideal configuration. I confess that I was lucky to be part of this group of pioneers.

Before concluding on this subject, it has to be underlined that, at the time of the G.I.E., Airbus had a limited number of aircraft types to sell and mainly to produce, as well a relatively modest number of customers (at least up to the mid-'90s). This could explain, maybe partially, the relative ease Airbus had to ensure very high-quality products and not limit the extent of support to the customers.

The arrival of the A320 family and its big success increased the volume of work at all levels and started imposing some rigid guidelines and processes, with the usual consequence of cost control.

In the first half of the '90s, Jean Pierson started doubting the continued viability of this structure and, in November 1994, publicly stated that:

"The G.I.E. had reached its genetic limits."

He started feeling the limits when a combination of factors started jeopardising his expansion plans and his attempts to improve the overall performances in terms of quality, lead time, and costs. This led to important decisions which transformed the company.

Before describing the second and third lives of Airbus, let me first relate the major things I lived during the group's pioneering era.

As Sales Director (1986–1988)

I joined the Africa sales region, under the leadership of the late Pierre Carpent, a former French Air Force pilot. Pierre was a very good boss, leading with fairness and humanity. We were a group of five sales directors—two French, one German, one Dutch, and me. We were working daily with marketing managers, contract negotiators, customer engineers, sales financiers, and customer support managers to run the sales campaigns.

We were in different offices, scattered around the building, and each function was reporting to other departments or directorates. I was lucky to find many known faces, in the different functions, thanks to my previous job at Tunis Air.

Some of them happened to work for the African region, and therefore, very quickly, a sort of informal team was created, grouping people like Henri de Sulzer, Georges Pons, and Jean Sutra, who all became some of my closest and most loyal friends at Airbus and who helped me create later the first commercial business unit.

As from the summer of 1986, I started travelling, mainly to Algeria, Egypt, and Morocco. Libya was in a stand-by situation due to political considerations, following all the US trade restrictions imposed on the country in 1978, 1981,

1982, and 1985. It took me some time to adjust to my new role, especially vis-à-vis the people I knew before.

Very quickly, two sales campaigns materialised: In Algeria, where the government gave the green light to the airline to buy three wide-body aircraft, and in Egypt, where the replacement of the old B737 and A300B4 fleets became a priority.

These two campaigns taught me a few key lessons about the reality of the airline world:

1. No airline, and in particular a state-owned airline, is immune to political influence.

2. Opportunities can arise from disasters.

3. In reality, a very limited number key people steer the course of a company or a country.

4. You know the true personality and the true values of a person in the middle of serious crises. The rest of the time, it is a piece of theatre, where everyone plays a role.

The Air Algérie campaign was initially a success for our A300-600 up to the negotiation of the contract, where a last-minute trick was thrown on the table. France had to confirm its agreement on certain conditions set for the sale of Algerian gas from Sonatrach (the Algerian national oil and gas company). We had no leverage on the French government and the concerned French companies to make them accept the set Algerian conditions.

A few months later, a contract for three B767-300 was signed with Boeing. God knows if the US government and the US companies accepted the same conditions (if they were even put on the table). We never had a full debrief, neither from the Algerian nor from the French authorities.

The discussions stopped overnight, with no further exchanges. A few months later, I went to visit Air Algérie and met the management. The whole previous negotiations were far away and somehow *forgotten*. The motto was, *Let us speak about the future*.

Air Algérie was a very good exercise to understand the complexity of the Franco-Algerian relations and to measure the limits of the margins of manoeuvre for the management of a state-owned airline. These limits were seen all over the MENA region, with the exception of one or two airlines.

Egyptair was a total success but with a very complex process controlled by the airline management. It was also a good case to demonstrate how a disaster can initiate a reconstruction programme. Egyptair expressed its desire to replace its old fleets of B737-200 and A300B4 with new aircraft. The A320 and the A300-600R were very serious contenders.

During the first half of 1987, several presentations and discussions took place, but the airline was not clear yet about the extent of the fleet renewal (seven narrow bodies and nine wide bodies) and was still debating about the main priorities. Suddenly, the airline faced the crash of an A300-B4 during a training flight at Luxor, killing five people.

Following this fatal accident, the chairman, late General Faheem Rayan, and his vice president of planning, Ismail Sherif, decided to make a complete review of the airline operation and undertake several reforms. I was present in Cairo at the time of the accident, and I was called to help with some interim lift.

Very quickly, the Tunis Air A300B4 FFCC was wet-leased (it means with crew and maintenance) for one month, and the airline shared with Airbus and Boeing some ideas on how to improve the airline operations. Thanks to the Tunis Air aircraft, Egyptair discovered that the Cairo/Paris/Cairo route could be flown with the same crew, without a night stop. Inspired by this experience, Egyptair renegotiated the agreement with the pilot union about time limitations and compensations.

A few months later, Egyptair cancelled the night stop in Paris and other destinations. But the most important, Egyptair came up with a complete fleet plan renewal covering several years. Based on this new fleet plan, we entered into a tough competition with Boeing and we engaged in lengthy negotiations with Egyptair.

After several months, we signed the A300-600R deal, and after close to one year and a half, we signed the A320 deal. In parallel, we conducted, jointly with Egyptair, several improvement programmes in maintenance and flight operations.

The A320 deal was a bumpy ride due to the Habsheim accident, which occurred on 26 June 1988. Airbus was obliged to organise a worldwide roadshow to explain the real reasons behind the crash, in the middle of a hostile environment due to strong lobbying—essentially by the French pilot union—to put all the blame on the aircraft and clear the crew of that demonstration flight organised by an airline (Air France).

Bernard Ziegler and I had the task to visit the main interested customers in the Middle East and explain to them, with facts, the circumstances of the accident. Egyptair was the top priority, followed by Royal Jordanian. The Egyptair A320 contract was at its last stage of negotiation, and we needed to urgently clear any suspicion about the safety of the aircraft.

Having had to wait for one and a half days for an appointment with Chairman Rayan, we finally got a slot at midnight. We took the opportunity of the available time to visit some historical spots in Cairo and concluded with a dinner in a sort of cabaret on the Nile (it was much more a restaurant with a small Oriental orchestra than a real cabaret with a show).

During this dinner, we made friendly encounters with two ladies, and one of them expressed her sympathy and sort of admiration for Captain Ziegler. They were seated next to our table and did not hesitate to start asking the usual questions:

"From what country? Why you are in Egypt? What do you do?"

We suspected they were professional entertainers and had to manoeuvre politely and diplomatically to leave at a relatively early hour to go to our meeting, without upsetting the two ladies.

In his book, *The Airbus Cowboys*, Bernard tells the full story with some variations.

The meeting with Chairman Rayan was surrealistic. It was in a small room next to his own office, under reduced lighting, with Captain Ziegler displaying charts and documents all over the place to explain the circumstances of the accident. The meeting lasted for more than two hours, and at the end, Chairman Rayan said he was convinced. The contract was eventually signed.

One of the surprising consequences of our long and rewarding campaign in Egypt was the friendship which developed between the Airbus team and the Boeing team, and in particular with Aziz Mohsen, the Egyptian sales Director of Boeing. Competing does not mean hating. Mutual respect is a must, and friendship does not mean betrayal.

My passion for history grew in Egypt, and I embarked my family on many trips to the main historical locations in the country. During one of our visits to Sakkara, I met an English gentleman who explained to me that the Egyptian history must be revisited in consideration of many new elements which were never included in the official narrative.

It took me some time to have a clearer understanding of what he meant, thanks to Sir Tim Clark, the current president of Emirates, who introduced me to a certain Graham Hancock and his theory about human history and the fact that the world is, in reality, older than what the official history is saying.

During close to three years of frequent presence in Egypt, I could appreciate the dedication of the management, their relative freedom of manoeuvre vis-à-vis the authorities, and the high level of confidence the President of the Republic and the Minister of Civil Aviation had in the persons of Chairman Rayan and Mr Ismail Sherif, the Vice President Corporate Planning.

I had a unique chance to interact with dedicated, straightforward, and transparent persons—certainly with a certain ego, but enough humility to accept advice and comments.

I could demonstrate to the management of Airbus that Egyptair could buy aircraft outside the complicated schemes the West thought were the rule for any business in the Middle East. Since then, Egyptair has had the full respect of all Airbus management, and the relations between the two companies became very strong.

One of the most salient characteristics of Chairman Rayan was his perfect knowledge of the Egyptian people. We were always impressed by his ability, at each delivery of a new aircraft, to entertain the crowd of Egyptian workers at the airport in a way it looked as if it was only for them, and he was the sole host of the show.

All the other guests on board the plane seemed to have disappeared, and it became a sort of communion between Chairman Rayan and the crowd. One day, he explained to me that the personnel of the company had to feel that they have the ownership of the assets, and this is why they have to celebrate the arrival of each new aircraft.

They must not appear as part of a show displayed for foreign guests but must feel that the show is solely for them. This was a unique case of a sort of populistic leadership, which proved to be efficient since Egyptair succeeded in carrying out many important reforms and improved a lot of its operations.

A surprising consequence of the Egyptair campaign was the visit I paid, as a member of the delegation led by Jean Pierson, by mid-1988, to the Minister-President of the Land of Bavaria, Mr Franz Josef Strauss, who was also the Chairman of the Supervisory Board of the G.I.E. Airbus Industrie. Mr Strauss was known as *the Bull of Bavaria*.

This visit was dictated by the need to seek the help of the German authorities to convince their credit agency, Hermes, to guarantee bank loans to Egyptair (we did the same with the French and the British). Mr Pierson was close to Mr Strauss, and he preferred to go through him to approach the federal authorities.

It was an intimidating exercise for me as a young salesman to summarise the Egyptair deal shortly and clearly in front of one of the most prominent political leaders in Germany. Luckily, he was fluent in English, and this made the exercise much easier for me. The only handicap I had during the meeting was the quantity of smoke generated by several smokers, mainly Strauss and Pierson, who were two heavy smokers.

It was a relatively short meeting of less than 40 minutes, during which I spoke for close to 10 minutes and had questions for another 10 minutes. Jean Pierson made me literally plunge into the cooking pot of Mr Strauss. Here I was, in front of the famous Bavarian Bull, who can do anything he decided. It was a great honour, but also a great responsibility. I needed to be precise, quick, and clear.

What was fantastic about this exceptional man was his ability to deal with politics and aeronautics at the same time. It was a parallel processing. Being himself, a pilot himself, Mr Strauss knew everything about aircraft, and this made him feared by the management of Airbus and of the German partners. He was, definitely, an exception to Felix Kracht's assessment of the Germans—the so-called *Bundesbedenkensordenträger*.

I was honoured and lucky to meet one of the pillars of the Airbus adventure. Let us not forget that F.J Strauss was the decisive hand in the creation of Airbus Industrie, thanks to his political influence in the Bundestag and his ability to provide the necessary financial backing to the project on the German side.

After we left, Jean Pierson looked at me and said:

"Not so bad, you passed the *bull test*, and you did not lose your balls."

The only problem is that Hermes will not move without ECGD (The British Credit Agency) and COFACE (the French Credit Agency) moving first, and this was far from being guaranteed.

Jean Pierson expressed his opinion in a simple manner:

"I trust Mr Strauss, but he cannot walk on water all the time."

He was right, and the Egyptair deal was finally financed by the Bank of New York.

Concerning Royal Air Maroc, all our sales campaigns were inconclusive, except for a marginal opportunity to sell four A321s, which were quickly transferred to another Moroccan airline before being sold (it was a short miraculous parenthesis). Despite my close and good relations with many executives in Royal Air Maroc since 1977, I have to confess my failure to convince them to buy Airbus aircraft. Here, I learnt another good lesson—*friendship is not enough to do business*.

As Regional Sales Vice President (1988–1998)

Very quickly, I got a confirmation that the existing sales organisation was not efficient. Each function involved in a sales campaign was reporting to a different department, and the sales directors, and even the regional vice presidents, had no real power on the teams.

This situation was further complicated by a silent opposition between the Sales Vice President and the Commercial Senior Vice President.

This antagonism affected the sales teams and led to a sort of failed *coup d'état* by the Sales Vice President in mid-October 1988. He had wrongly assumed the CEO would endorse his action, but it was exactly the opposite which happened. He left the company within a few hours, along with three other persons, including two heads of regions, and was replaced by my direct boss, the late Pierre Carpent.

At the same time, I was promoted to Vice President of Sales for Middle East and North Africa, excluding Tunisia. This promotion was decided, half an hour after the departure of the Sales Vice President.

The proposal was made to me by Stuart Iddles, the Senior Vice President of Commercials, by telephone. I was in Cairo, trying to close the negotiations of the A320 contract.

Stuart asked me a simple question:

"This morning, Jean Pierson asked the Sales Vice President to leave the company. I am restructuring the commercial directorate. Do you accept to take the position of Vice President of Sales, Middle East? I need your answer urgently because I have to send the new organisation to the executive committee and the Supervisory Board before the end of the day. Jean is aware of my proposal and is endorsing it."

I could only say yes. Two days later, I was in Toulouse, putting in place my new unit.

It is worth mentioning that Cairo and the Sheraton Heliopolis seemed to be the places where I learnt some of the most important news affecting my life.

On 7 November 1987, I was called by my brother-in-law to inform me about the coup organised by General Ben Ali to oust President Bourguiba after 31 years in power.

On 12 September 1988, I learnt the death of my mother, my sister-in-law, and my nephew in a car accident. My brother and my niece survived the car crash.

On 31 October 1988, I learnt that I was promoted, and this happened just a year and a half after joining Airbus.

As soon as I came to Toulouse, I asked both Jean Pierson and Stuart Iddles if I could implement some of the recommendations I had made in my famous report, in order to build my team. The answer was yes, and the first sales business unit was born. For the first time in Airbus, the head of a region had a controlling power over the members of the business unit: the salesmen, the contract negotiators, the marketing managers, the customer engineers, etc.

The Middle East team was built around the first core of people I mentioned previously and included people from diverse origins and backgrounds.

We had a former priest, a former rowdy, a former salesman of racket strings at Babola, a former Gurkha officer… We had people from different nationalities and, last but not least, we were the team with the biggest number of women in sales, marketing, and contract negotiation (the maximum was reached when the team was positioned in Dubai).

From day one, we decided to work as a unified team (I was not the boss: I was a coordinator and a facilitator). Each member of the team could replace another member for the basic duties (submitting an offer, collecting comments). This was the secret behind our motto: *constant presence in the field*. The customer was seeing frequently members of the Airbus Middle East team and got the impression that Airbus is constantly present.

Any important information was shared with all the concerned members of the team, and in case somebody made a mistake, all the team members were called for a meeting and jointly decided on the necessary actions. There was no place for a one-man show and for the prima donna spirit. In case of success, all the concerned members of the team shared the glory and the rewards, and not only the salesman.

All the team members were empowered and had the possibility to conclude deals when it was possible. There was no need to call me for a conclusion when it was not necessary. The presence on the photo of a contact signature was not a criterion for appreciation, reward, or promotion. Teamwork was the motto, and I am proud of having been part, for close to three decades, of a fantastic team.

Very quickly, the other heads of regions applied the same concept, and by mid-1989, the concept of the business units became a reality all over the commercial directorate.

In our business unit, we went further by introducing the Business Rationale (BR) to explain why we propose such a type of aircraft and such a number of units, with such a schedule, and what needs to be addressed in the offer to respond in the best possible way the customer requirements. We also introduced the Business Control Sheet (BCS), summarising all the commercial, technical, and financial aspects of the offer.

We had the first basic version of these documents, which we used for close to a decade. After some time, some of the other sales regions started using them. I generalised them and improved their content by late 1998.

Another major change introduced in the management of the Middle East region was that the priority is for self-sufficiency. By ensuring constant presence in the field, listening carefully to customer needs, and responding efficiently to them, we put in place a direct link with the airlines and minimised the need for political lobbying and the use of consultants and advisors. Within a few years, Airbus became part of the local landscape.

Thanks to a more widespread use of the Arab language (besides some members of the team having Arabic as a mother language, we had British, French, and Spanish citizens fluent in Arabic), we were able to speed up the flow of communication with the airlines and ensure a better understanding.

As from November 1988, I started visiting the airlines of the region, starting with Kuwait Airways, Saudia, Gulf Air, and the new start-up Emirates. These four airlines had plans to acquire aircraft within the next two to three years.

Thanks to my AACO connections, I knew a good part of the top management of the four airlines (even Emirates, because many key managers came from Gulf Air). Since Kuwait Airways and Saudia were already customers, I had to start with some support issues before discussing future plans. With Emirates, the ball was already rolling, and incremental orders were in preparation.

With Gulf Air, we had to go through a full introductory exercise. The creation of Emirates caused the departure of some key managers, and the newcomers were not familiar with Airbus. It took a few months before reaching the point where we started discussing the replacement of the B737 with the A320.

In 1989, we had a very busy year with several contract signatures and the start of many promising sales campaigns. But two events took the lead during 1989 and had a very important effect on the whole Middle East team.

The first event was the demonstration flight of an A310 in Iraq in May 1989. In the frame of a sales campaign of A310-300 aircraft for Iraqi Airways, we decided to take a Lufthansa A310, with a complete German crew complemented by our chief test pilot Bernard Ziegler and two other pilots. The Airbus delegation was important and was led by Stuart Iddles. I was part of the delegation.

I travelled to Baghdad before this trip, and I was impressed by the city and by the people. I found the Iraqi people well-educated and with a great sense of hospitality. The management of the airline was very professional and competent, but above all, it was intriguing by its large diversity.

During meetings with the executive committee, you found around the table a Sunni, a Shia, a Kurdish, an Armenian, an Orthodox, a Turkmen, and some other exotic communities like Nestorian or Babylonian.

This diversity was unique in the Middle East and in the world.

The Iraqi authorities organised a flight between Baghdad and Bassorah and allowed us to fly at low altitudes. We were all excited to visit Iraq and fly domestically. Many of us were curious to see the landscape between Baghdad and Bassorah from the air at a relatively low altitude. I was standing in the cockpit beside one of the close advisors to Saddam Hussain. Captain Ziegler was on the right seat, and the chief pilot of Iraqi Airways was on the left seat.

As soon as we reached the north of Nassirya, Captain Ziegler said to me in French slang:

"Mate à droite."

Which is similar to *having a right shufti*.

I turned my head and I saw this succession of crescents in the sand. Before I said anything, the advisor to Saddam Hussain told us in perfect French, with a slight Lebanese accent, that these were the famous bunkers. Ziegler and I were astonished by the quick reaction of the guy.

Our surprise was even bigger when he continued, in French, explaining to us that he was a fighter pilot trained in France, that his mother was Lebanese, and that the bunkers were built to be shelters for men and hardware, and that their disposition in the ground is random to minimise the effects of aircraft bombardment. This clear, unexpected explanation by a person close to Saddam confirmed to us that there was a big trust placed in France.

The scenery on our route was a magical landscape with a mixture of oases, deserts and lakes, and the simple fact that we were overflying the cradle of civilisations added to the magical aspect of this trip. We had a lavish reception at Bassorah airport, attended by all the high officials of the province and those who were on board. The demonstration was a success, and we signed a contract for five A310s one year later, just before the invasion of Kuwait.

The contract never materialised, but this trip and other subsequent trips to Baghdad had an indelible mark on my memory. I was not very much convinced that all the bunkers we saw could be busted in a few hours. I never visited Iraq after May 1990. I preferred to keep the positive clichés I had registered before the first Gulf War.

I never liked Saddam; he was the Stalin of the Arabs, but the country was a land of religious tolerance, culture, and very ancient civilisations. Iraq was relatively prosperous; the people were highly educated, and the standard of living was higher than that of most Arab countries.

It is a pity to see Iraq in its actual shape. Iraqi Airways was one of the best airlines in the Middle East and was playing the role of Emirates and Qatar Airways by offering connections from Europe to Asia.

In Iraq, we had our first experience with *espionage*. We had a warning from one of the European embassies to be careful about disclosing confidential information in taxis or in chauffeur-driven cars, to avoid leaving confidential documents unattended in the rooms, and to refrain from debating confidential items in closed areas.

After evaluating our respective locations in the Hotel Al Rasheed, we decided on a certain number of precautionary measures and decided to have our team meetings in the swimming pool, using a technique of clapping our hands in the water to jam any eavesdropping. It was funny to see the whole team coming to the swimming pool at the same time and staying for some time in the water, playing like kids.

One of the other important measures was to have one single copy of the offer or the contract, which we were taking with us all the time we were out of our rooms. The lessons learnt were used quite often, but we were not immune to sophisticated espionage.

There was, also, one very negative aspect of the Iraqi Pan-Arab dogma, and which I suffered from because of my Tunisian passport: it was the systematic political check after the normal immigration check at the airport when you leave the country. It was imposed on all citizens of Arab countries.

One has to go to a side office and get interrogated by an officer after getting the passport stamped by the regular immigration officer. This additional burden can last from a few minutes to much longer and was feared by a lot by Arab citizens. On the three occasions I travelled to Iraq, in addition to the A310 demo flight, I went through this side office.

It did not last long, but the questions were random and bizarre (in what city you are born? Since when you are in this job or position? Are you a member of a political party in your country?). I could avoid this question by just saying I was an immigrant in France).

A Tunisian friend who was the general manager of one of the major hotels in Baghdad explained to me, years later, the rationale behind this side office: it was the visible expression of the dictatorship.

But despite the negative aspects of the dictatorship of Saddam Hussein, nothing could justify or explain the invasion decided unilaterally by the USA in 2003. As far as the war to liberate Kuwait was justified, this one was a complete blunder and disaster, not to say a crime.

Destroying a country, a people, and a culture cannot be justified by a ridiculous rhetoric of destroying weapons of mass destruction (which did not exist) and establishing democracy (but at what price and to which end result?).

One of my main disappointments related to the Invasion of Iraq was to see many of my colleagues (mainly US citizens), who were supposed to be highly educated and exposed to the world's diversity, react like any basic redneck and show a very negative attitude vis-à-vis the Arab and Muslim worlds.

They refused any contradictory discussion, and I even saw some of my French colleagues engaging in harsh exchanges with our American colleagues. It was no surprise to me to see the same people supporting Trump later on.

I never suspected that populism could affect educated people holding executive positions in big corporations. It was a shock for me.

But this was an eye-opener, which showed me that I did not share the same values with them.

The second event was a visit to Beirut just a few days after the Taif Agreement, which put an end to the Lebanese civil war. In this instance, I landed in a country which had suffered a horrible war. I had been advised not to do the trip, but the chairman of MEA, late Selim Salaam, insisted and said he would personally guarantee my safety.

I had a warm welcome at the airport. I was taken immediately to the Hotel Summerland in West Beirut. The hotel was under the control of the Shia Al Amel Militia, under the leadership of Nabih Berri, the current speaker of the Lebanese parliament.

One thing was surrealistic in Beirut at that time was the ability of the MEA staff to move from any location to the airport in full security. The four bodyguards I had with me were all Shia, and they were telling me that they escorted technicians, crews, and cabin crews from different factions without any problem. MEA was acting as the sole common asset of the different factions, and nobody wanted to break this unifying symbol.

Here, an airline was the symbol of the unity of a fragmented country. Due to my full name, which is very common among Shia, I was treated like one of them and, surprisingly, I could access several places ravaged by the war, without any problem—neither from the Syrian checkpoints, the Christian checkpoints, nor from the Lebanese army checkpoints. The secret code was the name of MEA on the car.

I was impressed by these four guys from West Beirut, able to communicate and get safe access from other communities. I spent three memorable days in Beirut, where I enjoyed good restaurants and even did some golf practice with Chairman Salaam on the golf course next to the airport.

This was surrealist: was it true that a war had been raging in these suburbs? As much as I saw a lot of damage everywhere, I did not see anything on the golf course.

My visit was vital for Mr Salaam, who was keen to start rebuilding MEA and wanted to show the authorities that Airbus was supportive and ready to start helping at this early stage.

. One curious element of this trip was that I never had a written confirmation from the Airbus insurer that I was properly covered, but Jean Pierson had given me his word that he would ensure full coverage in case of any problem.

This was a strong commitment, and I fully trusted him, which was enough to make the trip possible. This strong gesture from Airbus was a fundamental element in the close relations established between the two companies.

The most striking image of this trip was a group of children playing in a huge hole (surely damaged caused by a bombshell) close to a Syrian checkpoint. Our car was queuing to be checked, and I had time to observe this surrealist scenery: five little boys, aged between eight and ten, were running up and down the little crater of 40/50 m in diameter, with two soldiers watching them as if they were parents watching their children playing.

At one moment, a soldier called the children and gave something to each one of them with a smile. What a paradox! It was a deadly but also a strange war. I had the impression that I was watching a war movie, but it was a sad reality.

On the last day of my visit, I had some free time in the late morning, and my security squad decided to take me for a tour in West Beirut, home of the Shia community. It was not far from the hotel and the airport. For more than two hours, I had a real immersion in another world—I felt like I was touring a film set of a *Mad Max* movie. I had a very warm welcome at all the shops I visited.

I believe my name and my nationality alleviated any possible rejection. On the contrary, my name and my family name, which, as I already said, were very common among the Shia community, generated a wave of sympathy, with the result being that I left the area with more than three kilograms of Lebanese sweets and a few souvenirs. Although these people were in dire straits, they did not forget hospitality.

When I came back to Toulouse, my colleagues did not believe what I told them.

J. Pierson was amazed by the golf practice and told me:

"Come on, you are adding a little bit."

I answered:

"Next time you see Chairman Salaam, ask him."

He effectively did when we met an MEA delegation in Paris a few months later to discuss the fleet plan.

Mr Salaam joked by answering in French:

"Everybody plays with bullets (in French *balles*). In Lebanon, we also have the right to play with balls (in French *balles*)."

I kept a sour taste about my visit to Beirut because I could not accept that the pearl of the Arab world became a field of ruins and that MEA, which used to be the best airline in the Arab world and the reference for everything (network, services, maintenance, flight operations, etc.), got into this situation. It has to be underlined that MEA helped many other airlines in the region to be set up and trained a big number of managers, pilots, and technicians.

Wherever I used to go in the region, I could hear people praising their instructors and/or coaches from MEA. I was impressed by one MEA captain who was a pilot instructor and who must have trained dozens of pilots and flight operations managers of many airlines (Egyptair, Saudia, Kuwait Airways, and others).

I was continuously hearing his name, all over the place:

"Captain Hanna."

He deserves a lot of respect and admiration.

Despite their outdated political system, based on a supposed balancing between the different religious groups, and the multiple mistakes made by the successive governments which opened wide the door for outside interference, the Lebanese people deserve respect for their resilience and their capacity to manage crisis with, very often, limited resources. They are real survivors.

In addition to these two special visits, I had a few more high-profile visits with the Airbus CEO in 1989, 1990, 1991, and 1992: to Amman, Bahrain, Cairo, Dubai, Jeddah, Sanaa (Yemen), and Kuwait City (in very special conditions). Each of these visits had its own flavour and focused on specific matters, but they had all in common meetings with the leadership of the country and lengthy and heavy meals.

In three of them, we had some memorable touristic excursions, particularly in Sanaa, where we had a constant come-and-go between the Middle Ages and modern days. We had one of the most enlightening political conversations with King Hussein in Jordan. In Dubai, we had a clear explanation of the long-term vision of the Emirate, and in Kuwait, we discovered the crazy consequences of the Iraqi invasion.

Some of the visits had long-lasting consequences on Airbus' presence and interactions in the Middle East.

Let us start with the visit to Dubai in 1989.

Jean Pierson knew and had already met the management of Emirates during the delivery of the first A310 on July 3, 1987. During its launch period, Emirates was followed by a team composed in its majority of British citizens led by Stuart Iddles, the Airbus SVP Commercial, with a very dedicated salesman called Stuart Wheeler.

These two persons were instrumental in securing the first purchase orders from Emirates for two A310s in 1985 and three A300-600s in 1988. When I took over the region, Stuart Iddles and Stuart Wheeler were very helpful in introducing me quickly to the Emirates management.

I discovered a group of motivated, dedicated, hard-working people believing in the future of their company. This group was led by Sheikh Ahmed Bin Saeed Al Maktoum, a high-ranked member of the Al Maktoum family, which is ruling Dubai. Sheikh Ahmed is the uncle of Sheikh Mohamed, the current ruler of Dubai, despite being younger than him. Sheikh Ahmed was and still is the chairman of Emirates.

He was then assisted by the late Sir Maurice Flanagan as managing director (this was his official title), and there were a group of young managers running the different departments like Sir Tim Clark for planning, the late Ahmed Al Mullah as head of line maintenance, complemented by some experienced guys like Captain Larry Smith for flight operations and Don Foster for on-board services.

Sheikh Ahmed impressed me with his human qualities (kindness, humility, accessibility, sense of humour, ability to show his friendship without inhibition or any protocol, etc.), and his solid knowledge of aviation.

It is a great pleasure to discuss with him in Arabic and cover multiple subjects exceeding the sole scope of aviation. Over time, we developed a strong and friendly relationship, and he became close to my family. I believe this strong relationship helped me in dealing with Emirates and in deciding later to move to Dubai.

Jean Pierson decided to come and see how this young airline was changing the rules of the game in terms of new standards of services, especially, after receiving their first A310s equipped with Satcom, individual videos, and more comfortable seats.

The main event of the visit, which took place in early 1989, was the meeting with Sheikh Mohamed Ben Rashid Al Maktoum, Minister of Defence of the United Arab Emirates, and the de facto the man in charge of implementing the Dubai vision for the future.

During this meeting, which lasted for close to one hour, in the presence of Sheikh Ahmed, Brigadier Barclay (advisor to Sheikh Mohamed), Maurice Flanagan, Tim Clark, Stuart Wheeler, and myself, Sheikh Mohamed gave us a comprehensive lecture about his vision for Dubai and explained why and how the city would become a double world hub: a maritime one, thanks to Jabel Ali port and its associated free zone, and an aviation one, thanks to the airport and its planned associated free zone (he mentioned there was a study about an aviation free zone).

He gave some gross figures about the tonnage of freight, passengers in transit, and tourists. He was passionate in his presentation and confident in the achievement of the goals. He put some focus on Emirates and its plans and showed full knowledge of the operation of an airline.

At the end of the meeting, and while we were leaving the lounge where Sheikh Mohamed received us, Brigadier Barclay took Jean Pierson on the side and insisted that Airbus must believe in the vision of Sheikh Mohamed and support Emirates. In particular, he strongly insisted that the planned size of the fleet by the year 2000, estimated at around 50 airliners, would be achieved.

During the lunch which followed the meeting, hosted by Sheikh Ahmed and attended by Maurice Flanagan and other top managers of Emirates, both Sheikh Ahmed and Flanagan insisted again on the viability of Emirates' ambitious plan.

A few months later, Dubai organised its first airshow, which was attended by a big delegation from Airbus, led by Jean Pierson. During a ceremony at the airshow, we signed a contract for some additional aircraft.

Sheikh Mohamed attended the signature and did not forget to remind Jean Pierson about his plans, by telling him:

"You see, we continue to order additional aircraft."

Emirates reached a fleet of A300-600s and A310s totalling close to 20 units within a few years.

Until his retirement in March 1998, Jean Pierson kept visiting Dubai and the Emirates on a regular basis and developed strong relations with Sheikh Ahmed, Maurice Flanagan, and Tim Clark.

The other important visit was to Amman, Jordan.

I knew Royal Jordanian and their President (at that time), Mr Ali Ghandour, from my days with Tunis Air. The airline had a good reputation for dynamism and good service. They seemed to be the successors of MEA.

Another important point for Royal Jordanian was the strong and continuous support the late King Hussein was showing to the airline. He was himself a good pilot and was perfectly knowledgeable of all aspects of aviation. With Jean Pierson, we decided to go to Amman to unblock some dragging discussions about the financing of the A310s and the finalisation of the A320 deals.

I discovered that Jean Pierson had already met King Hussein on at least two occasions and that they developed a friendly relationship. After one meeting at the Royal Jordanian (known as Alia at that time) headquarters, Ali Ghandour told Pierson that the King had accepted to receive him at the palace in the afternoon.

Jean Pierson was supposed to meet the king alone with Ali Ghandour. At the end of the lunch, which followed the morning meeting, and while we were going to take a car for the hotel, Ali Ghandour stopped us and said that the Airbus customer financing director, who was part of the delegation, and I should come along with him and Jean Pierson to the Palace, and that we were to wait in a separate room, because the King could call us for some details.

At around 4 pm, Jean Pierson and Ali Ghandour were taken to meet the King, while my colleague and I stayed in the waiting lounge, with enough juices, water, and sweets to feed several persons for a few days.

After some 40 minutes, Ali Ghandour came to take us to join the royal meeting. From his moves and his behaviour, you could see that he was familiar with the place.

The King stood up and greeted us warmly in perfect English, adding in French:

"Bienvenue."

Ali Ghandour mentioned that I was Tunisian, and the King switched to Arabic and said:

"Assalama," which was the Tunisian expression for welcome.

I was impressed by the simplicity and politeness of King Hussein. He focused on making us feel comfortable before starting to ask any questions. Jean Pierson introduced both of us to the King, with a short joke about our common Tunisian background. The king was so nice that he took a few minutes to ask questions about the major differences between Bizerte and Sousse.

After this friendly and funny preamble, the King asked us some very specific questions about many aspects of the contracts.

He listened carefully to our explanations, and then he started giving us a sort of overview of the political situation in the region, its effect on Jordan, and the challenge met by his country to ensure the development and the welfare of its people while at the same time spending money on ensuring a proper capacity of defence against any aggression.

He underlined the need for innovative financing to bridge the gap between necessary and available funds. He stressed that aviation was a priority and that Alia must play a leading role in the region. The most important part of his briefing was the description of the regional geopolitical situation and the role of each country.

He was very visionary in his assessment of the situation in Iraq, and the possible consequences of the cooling relations of this country with some of its Gulf neighbours. He also insisted on the fact that Jordan needed to see the Palestinian problem solved with a real peace accord between the Israelis and the Palestinians. This will help strengthen the stability of the country and boost its economy.

He also gave us his opinion on the Iranian effect on the region after the revolution. I found his judgement very accurate and balanced. There was a destabilising element in the Iranian revolution, but the Arab countries must avoid putting full responsibility on Iran for their problems, while some of them had their origin in purely internal causes with no link to Iran.

I was surprised by the openness and the desire of the King to express his views in front of foreigners. It was a lesson in geopolitics which made me feel humble in my pretention to understand the ever-changing Middle East political landscape.

After finishing the meeting, the King took a few minutes to ask me some questions about Tunisia, concluding by telling me he hoped the best for my country.

Because of this discussion with King Hussein, we were obliged to find an innovative solution to finance or refinance the A310s of Royal Jordanian.

Royal Jordanian continued to suffer from some financial shortfall for a certain time, despite the many efforts by the management and the authorities to alleviate the burden. It was true that renewing the entire fleet within a relatively short period of time and developing the technical and training capabilities in-house were very costly exercises.

For this purpose, and sometime later, Airbus participated in the creation of Waha Leasing (or Oasis Leasing), with other French companies (Aérospatiale, Thales... which had some offset obligations vis-à-vis the authorities of Abu Dhabi) and the support of Banque Indo-Suez, particularly its representative in the Middle East, the late Mrs Anne-Marie Siffroy-Pitlack, and her deputy, Seethor Senghor.

To achieve this goal, I teamed with my colleague, Olivier Gain, from our Customer Finance Department to discuss the project with Dr Amin, Director of the Offset Office of Abu Dhabi (ancestor of today's Mubadala).

On top of these discussions, I had several meetings with His Highness Sheikh Mohammed Bin Zayed Al Nahyan, who was in his mid-30s and cumulated the role of chief of staff of the UAE army and the role of President of the Offset Office, plus other responsibilities.

Waha Aircraft Leasing entered a sale and leaseback transaction with Royal Jordanian for their fleet of A310s. This was the first nucleus of the future Waha Leasing Company and, later on, of Abu Dhabi Waha Capital.

On this occasion, Airbus solved the problem of Royal Jordanian and helped the authorities of Abu Dhabi in creating a leasing company without having any business with Abu Dhabi (except two A300-600 VIP aircraft sold a few years earlier) and without having any obligation of offset.

I had several other opportunities to meet Sheikh Mohamed Bin Zayed during my career at Airbus. This enabled me to better know the person, understand his vision for Abu Dhabi and the UAE, and appreciate his qualities and unique personality.

From my first encounter with Sheikh Mohamed, I discovered a man who had a vision, knew where to go, and was very keen to go step by step and involve the people around him. I remember him explaining the role of the Offset Bureau and what he was expecting to achieve with it.

He was modest in his immediate objectives, but one could see that his long-term plan was ambitious. I was not surprised when Mubadala was created as a natural extension of the Offset Bureau (the word Mubadala means *exchange* or *offset* in Arabic).

During all my meetings with Sheikh Mohamed, I had in front of me a soft speaker, keen to explain quietly and precisely what he wanted you to grasp. I never heard him bragging or menacing.

For close to 30 years, I was a privileged witness to the rise of two prominent personalities in the Gulf: Sheikh Mohamed Bin Zayed and Sheikh Mohamed Bin Rashid. A lot of my achievements in aviation are in connection with Abu Dhabi, Dubai, and the UAE.

I had the opportunity to meet King Hussein on two other occasions: an AACO general assembly in Amman and an A340 demonstration flight at Zarqa Air Base.

The demonstration flight was an opportunity to appreciate the sense of humour of the King, despite stressful circumstances. The event took place the next day following a terrorist attack on the old Roman Amphitheatre of Amman, which left some casualties, including Israelis.

The King, who was planning to fly the aircraft, came late to the airbase and was not relaxed at all. He explained to Jean Pierson (who was keen to be part of the trip) and to our Chief Test Pilot, Captain Ziegler, the reasons for his delay and visible stress.

I succeeded in squeezing myself into a standing position in the cockpit, thanks to my strong friendship with Captain Ziegler. We had developed a sort of routine following the numerous demonstration flights we organised together in the region.

During this flight, the new President of Royal Jordanian was keen to sit next to Jean Pierson in the cabin and declined to come into the cockpit, leaving the proposed jump seat to the airline chief pilot. Therefore, we were five in the cockpit.

The King started flying with some hesitation and lack of focus until Captain Ziegler told him:

"Majesty, the day you get fed up with your job, I offer you to join us as a test pilot."

The king laughed and, within seconds, took full control of the flight, calling Air Traffic Control to request authorisation to fly at a low altitude above Amman. He performed an immaculate low-level flight pass with some daring military manoeuvres. All the passengers were bluffed, and the landing saw sincere and loud applause.

With Jean Pierson, we came quickly down the stairs to greet the King when he left the aircraft.

He stopped at the aircraft door with Captain Ziegler, exchanged a few words, and then ran down the stairs. He shook hands with Jean Pierson, and then it was my turn to shake his hand. Suddenly, he kept my hand in his and looked up to Captain Ziegler:

"I do not forget your offer. We need to discuss the details."

Only those who were in the cockpit understood the King's comment, and I had to brief Jean Pierson about their meaning. But before I finished my briefing, King Hussein came back towards us and asked both of us to follow him to a yellow Mercedes car, which was parked less than two hundred metres from the aircraft.

During the walk, he related his exchange with Captain Ziegler and told Pierson that he is taking him to the officers' club at the Airbase for a drink.

The King sat in the driver's seat and invited Jean Pierson to sit in the passenger front seat. Two bodyguards were sitting in the back. I was closing the door on Jean Pierson's side when he stopped me and showed me a Kalashnikov or something similar, blocking him on his left leg.

The King noticed Jean's reaction and said to me in Arabic:

"He can put in the front of the seat between his legs. There is no risk because it is in the safety position."

But Jean, maybe feeling at ease in the presence of the King, showed me his precious part and said to me in French:

"Is there no risk for this?"

The King understood the situation, laughed, and said to Jean Pierson:

"You have my guarantee, but to make you feel more comfortable, I'll put it on my side."

On this funny exchange, I closed the door, and the King drove the car to go to the officers' club.

This demonstration flight remained one of the most iconic ones, and King Hussein became very popular within Airbus.

The *Apocalypse Now* visit to Kuwait City.

In the spring of 1991, we decided with Jean Pierson to take the full Airbus team who was involved in the sales campaign to Kuwait Airways, to Kuwait to try to close a deal. This visit was agreed upon with the top management of Kuwait Airways, and a meeting with the Crown Prince was planned.

Flashback

Following the invasion of Kuwait by the forces of Saddam Hussein, one part of the Kuwait Airways management, under the leadership of Chairman Ahmed Al Mishari, took refuge in Cairo, where they set up a temporary headquarters, while the other part, who remained in Kuwait under the leadership of Director General Ahmed Al Zabin, had a very limited capacity of operation and was a simple source of information.

The main problem faced by Kuwait Airways was the illegal seizure of their fleet which was parked at Kuwait City airport by the Iraqi army. Kuwait Airways had to prove its ownership of this fleet to the United Nations and to different international institutions, with the purpose of recovering the fleet or to be compensated, in case of a total loss. In addition, there was a need for help to be able to operate as Kuwait Airways from remote locations like Bahrein and Cairo.

As of August 6, my team (Gerry Sharp, Henri de Sulzer, and Georges Pons) and I joined the nucleus of Kuwait Airways management, first in Bahrain for a few days and then in Cairo. Very quickly, we started collecting all the necessary documents and putting together the required files for the proof of ownership of the seized fleet.

Within a few weeks, we succeeded in having everything ready and given to the Kuwaiti legal authorities, based in the city of Taif, in Saudi Arabia. This joint effort created strong bonds between the Airbus team and the Kuwait Airways team. We were following closely the build-up of the international coalition to liberate Kuwait, either from Bahrain, Dhahran, or Riyadh, while we were preparing a new fleet plan for Kuwait Airways post-liberation.

By mid-January 1991, the fleet plan was ready, and Kuwait Airways started approaching Airbus and Boeing for package proposals for 16 aircraft. While the troops were fighting on the ground, we were fighting against Boeing in the Kuwait Airways Cairo offices. During the close-to-one-month war, we were able to follow the operations either in Bahrain or in Riyadh. We saw the British Tornadoes taking off from Bahrain to bombard Iraq.

We experienced alerts for Scud attacks, with, on one occasion, a missile hitting a spot situated a few hundred metres from our hotel (luckily without serious damages), and we had multiple opportunities to sit and discuss with politicians and military personnel from the coalition countries.

These discussions were very interesting because they gave us an insight into the progress of the war and some ideas about the real perception each country had of the invasion of Kuwait and of the consequences of the war. It was not surprising to detect slight differences between the different countries.

From all our discussions, one thing became obvious: the coalition did not have a clear idea of the future of Iraq and of Saddam and did not grasp the strategic importance of a stable and strong Iraq. The future confirmed this impression. Not surprisingly, the British had the most realistic and balanced view.

By the end of March, we entered the critical phase of negotiations while the Kuwait Airways top management started transferring back the operations to

Kuwait City. This move was difficult, but it was necessary to show that Kuwait was still alive and kicking. To show our support to the Kuwaitis, we decided to travel to Kuwait.

First, Gerry Sharp succeeded in convincing the British authorities in Bahrain to have a seat on a Hercules flight of the British Royal Air Force (RAF) from Bahrain to Kuwait and back. He came back within two days. He described what he saw on the spot, and we were all impressed. Then we planned our high-level trip with a private jet.

We entered the Kuwaiti airspace in the middle of the afternoon, but we had the impression it was night time due to the black smoke emitted by the burning oil wells. From the plane, we had a view like the one in the movie *Apocalypse Now*, where a helicopter was overflying a devastated battlefield, with fire and smoke everywhere. We were just missing the music of the *Valkyries*.

After landing, we discovered a contrasting situation: the city was not too much damaged, but it was emitting a strange feeling of desolation and sadness. We went to the Le Meridien Hotel, located downtown. The rooms were in good shape, and everything was functioning. The next day, we had a formal negotiation meeting with Kuwait Airways in a prefab meeting room annexed to their headquarters, which were not fully usable.

After several negotiation sessions in Cairo and in London, some at the level of the two CEOs, we were reaching the limits of our possibilities. Consequently, it was difficult to give sizeable additional concessions to the Kuwaitis, and this was a source of frustration.

We knew that Boeing was using all the US government clout to impose their offer, but we could feel that the Kuwait Airways management was somehow grateful for all our help during the times they were in dire straits.

The meeting with the Crown Prince was scheduled for the next morning. We prepared a list of bullet points for Jean Pierson with a short summary for each point. Among the list, I proposed to explain to the Prince that the Kuwaiti Investment Authority (KIA) owns around 18% of Daimler, which, in the meantime, was controlling 100% of DASA, which owned 31.4% of Airbus Industries.

Therefore, *de facto*, Kuwait owns around 5.6% of Airbus. I knew that the relationship was slightly more complicated in terms of direct revenue (mainly due to the G.I.E structure of Airbus), but it was an interesting approach to show that the two companies were sort of cousins.

The explanation was given to the Crown Prince, who was impressed and asked the Minister of Finance, who was present at the meeting, for confirmation.

There was a big laugh, followed by a logical comment:

"Yes, if DASA makes money and if Daimler makes money."

Nevertheless, the concept of a close relationship was admitted. The main convincing elements in favour of our offer were the lower total cost (disclosed by Kuwait Airways years later) and the earlier delivery dates (known from the beginning). When leaving the meeting, Jean Pierson was, with good reasons, very optimistic—after a few more months, we could sign the formal contract for the purchase of the aircraft.

Two lessons learnt from the Kuwaiti deal: one is the importance of supporting the customer during difficult circumstances and to be constantly present in the field, and the other is to never give up, even in front of strong political lobbying supporting the competition.

This deal with Kuwait Airways was finalised a few months later for 16 aircraft of different types. Among the commitments of Airbus towards Kuwait Airways was the assistance to set up a leasing company.

In this respect, with the help of a small group of people from Airbus, I coordinated the actions of assistance to the Kuwaiti team in charge of the creation of ALAFCO, the Kuwaiti Leasing Company (establishing the business plan, defining the structure, creating the logo, to name just a few activities).

But I kept a sour taste of our Kuwait Airways sales campaign because it was a source of harsh and tricky attacks against Airbus and myself by certain political groups in the country and by certain local and even international media. I understood the continuous confrontations between the successive parliaments and the successive governments and the never-ending accusations between the two sides.

But I never suspected that Airbus would be used one day as an excuse for a fierce battle, which lasted for some time. Many years later, I had a frank discussion with some high-ranked officials about what happened, and I got a fair and frank explanation, as well as a sort of apology.

Therefore, Airbus contributed to the creation of two leasing companies in the Middle East: Waha Leasing and ALAFCO, which grew and became prosperous with important investments and diversified portfolios. Both are still running, more than 30 years after their creation.

Another special purpose vehicle, called SAMA, was created to refinance the B767 that had been bought back from Gulf Air, and which was on Japanese tax lease. The chosen scheme was very sophisticated and innovative and drew a lot of congratulation from the financial community. SAMA had a limited life, due to the structure of the financing.

This sophisticated transaction was part of the agreement signed with Gulf Air for the sale of several A340-300s, in 1992. One of the most interesting aspects of the negotiations of this contract was the process followed by Gulf Air.

All the competitors for the airframe (Airbus, Boeing, and McDonnell Douglas) as well for the engines (General Electric, Pratt & Whitney, Rolls Royce, and CFMI, the joint venture between GE and SNECMA, now SAFRAN, set to build the engines for the A320, A340, and B737) were called to Salalah, a city in the southeast of Oman, in the same hotel where the board of Gulf Air and the evaluation committee were meeting.

During one full week, each manufacturer was called in several times to face a sort of *tribunal* composed of the board members and the evaluation committee members, and had to present his offer, clarify some details, and answer questions. One of the most discussed points with the Gulf Air committee was Airbus' ability to achieve the financial scheme proposed for the B767.

I remember the negative comment by the CFO of the company who said to us:

"No way, Jose, you cannot apply your scheme for a Japanese tax lease."

Despite the clear explanations of Paul Meijers, our Dutch aircraft financing director, the guy did not change his mind. It took us a follow-up meeting, in Bahrain, a few days before the final decision, with all the supporting documents, to convince him.

Between two sessions in front of the *Tribunal*, we were all together, chatting, eating, playing chess or cards, swimming (only in the swimming pool because it was the monsoon period, and the sea was very dangerous), or going to the gym. There was no competitive spirit, we were simple colleagues trying to spend the time the best way possible. The Airbus and Boeing teams knew each other, and we had no problem to mingle and to have funny discussions.

We were also close to the CFMI and GE teams. The striking element was the reluctance of the McDonnell Douglas people to be friendly with the rest of the competitors. With my team, we had no problem to share a table for lunch or dinner with the Boeing team or any engine manufacturer team, but we could not do it, even once, with the McDonnel Douglas guys. It seemed they were resigned to their inability to compete against Airbus and Boeing.

Or maybe, they had a culture of suspicion and mistrust in any circumstance. Anyway, with the Boeing team and in particular their sales director for Gulf Air, Mohamed Abdelbari, we had memorable moments of laughter when we started exchanging North African and Egyptian jokes. Mohamed was a long-time friend. After a week full of stress and fun, we left Salalah and waited for the verdict which came a few days later in our favour.

One of the key elements in the success of our offer was the buy-back of the B767s. The first A340 was delivered in 1994. On this occasion and for the first time, the delivery lunch or (dinner) was organised in my house, in a form of a garden party, for around a hundred persons, with the main attraction being a pétanque competition, won by Jean Pierson.

Not a surprise as Jean, as a typical Mediterranean, loved this metallic ball game played on the French shores of the Mediterranean while enjoying a good *Pastis* aperitif drink.

This game would become his trademark in many future events. Pétanque is a game played with metallic balls, thrown by the players with the aim to get the maximum number of balls at the shortest distance from a small wooden ball called *cochonnet* or piglet, which is thrown at the beginning of the game to play the role of the target. Pétanque can be played between individuals or teams of two or three. Each player has three balls at his disposal.

Gulf Air inaugurated its first non-stop flight to New York the same year. I was invited to the inaugural flight and experienced, for the first time, a flight duration of more than 14 hours. It was a real game changer for the connections between the Gulf and the US. We enjoyed our few days stay in New York, and particularly a succulent dinner in an empty Italian restaurant. The ambience was bizarre; however, and we discovered later that it was owned by the mafia.

Today, many airlines fly non-stop from the Gulf to the US. Emirates followed with the A340-500 and then with the B777 and the A380. Etihad and Qatar Airways did the same, and there are currently over 10 of daily connections between the main Gulf cities and several cities in the US.

Salalah will remain a nice souvenir for me for several reasons. First, we won a very tough competition. Second, I had the opportunity to discover the beautiful region of Dhofar with its cool weather and its landscape closer to that of the Swiss Alps than to that of the Gulf region.

Third, I also discovered some intriguing historical sites like the supposed tomb of Prophet Job, thanks to visits organised by the Gulf Air management; and fourth, I enjoyed an unparalleled Arab hospitality. While all the competitors were paying for their rooms, they did not pay anything for the food, from breakfast to dinner, this in addition to one gala dinner organised the last day. The food was excellent and in quantities I'll never forget.

I became fond of Oman, for the friendly and peaceful behaviour of its people, for its culture, for its beautiful architecture which looks like the one described in Sinbad books, for its attractive historical heritage and the beauty and variety of its landscapes which range from arid desert to green and rainy mountains.

It must be underlined that our success was due to our strong points in finance: be creative, while abiding to all legal constraints, in putting in place innovative financing packages. Thanks to a motivated customer financing team led initially by Laurence Barron and subsequently by Benoit Debains, and comprising a group of sharp experts like Yann Ballet, Nigel Taylor, Paul Meijers and many others, there was no difficult case for which we could not find a solution.

Another achievement deserves to be mentioned, for its specific conditions and features: the sale of two A300-600R to Iran Air. On 3 July 1988, the US caused an aviation tragedy in the Arab/Persian Gulf by downing an Iran Air A300-B4K, flying between Bandar Abbas and Dubai, with a missile fired by the US Navy from USS Vincennes. 290 passengers, including 66 children, were killed.

Following a denial, and then the confession of an error due to the confusion of the radar signature of the Airbus aircraft with that of an Iranian F14 Tomcat, the US government expressed regrets and, sometime later, allowed Iran Air to acquire two A300-600R aircraft from Airbus. This was a zero effort in terms of sale, but a huge mountain in terms of finalising the sale and particularly for the payment terms.

Thanks to our sales financing team, a shrewd solution was found, in full respect of the US embargo, the Iranian constraints, the European banks' limitations in dealing with Iran, and, of course, the Airbus interest to be paid on delivery. Finally, all went well, and we delivered the first aircraft in 1994.

This first delivery was memorable for all the Airbus delegation led by Jean Pierson.

Personally, it was my first encounter with the Tomcats of the Iranian Air Force, which came to escort the new aircraft upon its entry into the Iranian airspace. It was an impressive view, having these iconic machines, which were not seen in any foreign country outside the US, except in Iran, which bought around 79 units of the model during the Shah era.

Having four Tomcats escorting our aircraft was something new for me and for most of the Airbus employees who were on board. Jean Pierson was curious to know how the Iranian could maintain such sophisticated aircraft, despite the embargo.

We had a very vague answer from the chairman of Iran Air and a strange answer from a high-ranked officer from the Air Force, who simply said:

"We build the parts!"

I had already visited Tehran before this delivery, to discuss the contract and the different conditions of the deal, but the actual visit, on the delivery, will remain unique. The hospitality, the sites visited, and the quality of the discussions we had with people at the highest level of the hierarchy of the country. I was surprised by a relatively frank and open dialogue I personally had with officers, clerics, ministers, etc.

It seemed there were instructions to our Iranian counterparts to be more open and speak about everything with their guests. Somehow, we were the friends of Iran, and we were bringing a high-tech aircraft to their country. The only things which were not negotiable were the sites we had to visit.

It was, clearly, a list defined by the leadership of the country to show us the greatness of the Iranian civilisation, the supposed benefits of the 1979 revolution, and the vital role of the supreme leader, Ayatollah Khomeiny.

Two things must be underlined. First, it was the exceptional hospitality we enjoyed during our stay, in all aspects; and the second, the active role of women in the society, despite the fundamentalist regime. We saw women in the airline and some of them at managerial positions. We saw women at bank desks, as bus drivers, as stewardesses, as airport employees, etc.

This was a stark contrast with Saudi Arabia. I could have a confirmation of both impressions during my subsequent visits to the country in the following years. Recently, things are also changing in Saudi Arabia.

One important element allowed me to always have a very warm welcome from the Iranians. It was my full name, Mohamed Habib Fekih (which in Arabic and Farsi is written Al Faqih). First, Mohamed Habib is a favoured first name for Shia people, and second, my family name is identical to the title of the

Supreme Religious Leader, who is the real ruler (Vilayet Al Faqih, or rule of the religion expert).

In addition, the fact that Tunisia was the first country where the Shia followers established the first Shia state in the world, well before Iran (the Fatimid dynasty, which controlled most of the Arab world, built Cairo and Al Azhar University), in the 10th century, added to the warmth of their hospitality.

At each visit, I had questions about the Fatimids and their capital Mahdia, which is 60 km from my birthplace. Sometimes, apparent small details can play an important role in shaping relations between people. Here I can say I was simply lucky.

One last and important lesson I learnt from Iran was how interwoven civilisations and cultures are. Here is a country which most of the world population considers as Oriental and Muslim, and which is very often considered as part of the Arab world, while it—de facto—is much closer to Europe than people assume.

First, Farsi is a language which is closer to German or Dutch than many other European languages (I was surprised to read, written in Arabic letters, on the disembarkation card the word *Name* as pronounced in German, to designate the family name. More surprisingly, you can hear an Iranian presenting his brother, sister, mother, father, and daughter with similar words to the German expressions, and praying to God, pronouncing the word like in Dutch).

In addition, many customs and uses are close to what could be seen in the Germanic world, not so long ago. The name Iran, which started to be used from the 19th century to replace Persia and became the official name of the country as of 1935, means the land of the Aryans, and the title of the king *Shah Aryamehr* could be translated to King Light of the Aryans.

This strong link with the Germanic world is ignored in the West, and a lot of misunderstanding related to Iran could be linked to this ignorance. To analyse Iran, as merely a Muslim country is not sufficient to allow a full understanding of its complexity. I personally enlarged the discussion to these aspects with my Iranian counterparts, and it helped me learn a lot of things.

Last but not the least about Iran, I became aware of a supposed clash between Sunna and Shia long after the Iranian revolution. In my home country and in all of North Africa, we never felt there was a fight between the two branches of Islam. We knew these existed, that is. I have the impression that the supposed clash started resurfacing after the Iranian revolution and was progressively promoted by people outside the Muslim world.

It is true that some sporadic problems arose over the last 1400 years, but their frequency and intensity had sharply declined during the last five centuries. The future will tell if this provoked clash continues or if it will backfire against its promotors.

In 2015, I had the opportunity to travel again to Iran, with respectively the, at the time, CEO of Airbus, Tom Enders, and the COO, Fabrice Brégier, in the frame of the Joint Comprehensive Plan of Action (JCPoA) which was signed, between Iran and a group of countries comprising China, the European Union, France, Germany, Russia, the United Kingdom, and the USA, defining how to control the Iranian nuclear programme and how to lift progressively the sanctions imposed on Iran. The purposes of these visits were multiple.

First, it was to agree on the main items which needed to be part of any possible aircraft purchase agreement (like *snapback* or return to square one in case Iran does not fulfil its commitments, which caused real headaches during the negotiations). Airbus and Boeing were allowed to sell, jointly, more than 200 commercial airliners to Iran, and both companies engaged in serious discussions with Iran Air and other authorised airlines in Iran.

The negotiation process was very complex because each clause had to be approved by a group of lawyers in charge of checking the conformity with the JCPoA.

Second, to discuss all the areas related to the technical support and to the training of the personnel.

These visits were very interesting, because they took place during a period of opening towards the West. Therefore, we had the opportunity to discuss more openly with our Iranian counterparts and to visit places which were not allowed previously, like the Museum of Modern Arts, which contains a big number of works of Andy Warhol (it would surprise many people to learn that an important part of Warhol's work is in Tehran and not in New York).

The big contract with Iran Air was signed by the two CEOs of Airbus and Iran Air at the Élysée Palace in Paris in the presence of both presidents, François Hollande of France and Hassan Rohani. I was happy of the positive conclusion and optimistic that we were entering a period of peace and prosperity.

But I was puzzled and concerned about this famous clause of *snapback*, which could bring back everything to square one. I never expected that the

President of one of the signing countries would simply and unilaterally decide to get out of the deal.

I could witness the delivery of the first three (and only) new aircraft to Iran Air: one A321 and two A330-200, in the first quarter of 2017.

What a waste of time and energy by dozens of people from both sides.

In the first years of the '90s, my team had a succession of positive achievements, with a big number of contracts signed with several airlines in the region (Al Yemda-South Yemen, Egyptair, Emirates, Gulf Air, Kuwait Airways, MEA, Royal Jordanian, Yemenia-North Yemen), plus some aircraft for private operators. Consequently, the Airbus market share in the region made a huge jump, reaching close to 65%, after hovering for years at less than 10%.

However, we were still lacking some of the *majors* of the time. Here below some examples of how things can turn out positive or negative, and which illustrate the need to be persistent and to never give up despite your wish to throw the towel and forget about that particular airline. Persistence and patience are really a must in that job.

At the time, there was a major target which we needed to achieve. It was Saudia Airlines. For this purpose, we had a unique opportunity to return to the airline, after close to 10 years after the delivery of the last of the 11 A300-600 aircraft ordered in the early '80s. The airline had issued a call for tender for 29 single-aisle (or narrow-body) aircraft and 23 wide-body aircraft, in late 1992.

We worked hard to build a very attractive offer with a lot of incentives, particularly in terms of financing, where we were fully aware of the sharp shrinking of the Saudi revenues, due to the fall in the price of the oil barrel ($11).

Above all, at that time, we had the best combination of aircraft between the A320 and the A340 thanks to their unique cockpit commonality, but also because the airline was envisaging to use the wide body on very long-haul thin routes, which the A340 was the first ever aircraft to be designed for.

The sales campaign was very hard, with a very strong lobbying from the US government. We were very naïve, and we counted on our own efforts and on some very polite diplomatic messages from the four ambassadors of our partner countries.

We enjoyed superb lunches and dinners with the ambassadors and their economical attachés, during which we passed very elaborate messages to be used to defend our case. The two most active ambassadors were the French (who is very intellectual and writes books) and mainly the British (who is more business minded).

We started facing serious challenges when the airline management started criticising us for the way we were conducting our campaign.

It was true that two British senior members of British Aerospace had been in contact with some Saudi officials during some government visits and some discussions about the then Al Yamama project, which was initiated in 1985 by the sale of 96 Tornado aircraft by British Aerospace to the Saudi government. I believed they had passed our marketing messages to the Saudi authorities.

In front of this criticism, I asked my management to let my team be the only one to deal in the future with the sales campaign and to inform us of any political support or marketing support from the partners, to be able to answer any criticism. This was true for several months until we started having some positive signs that our offer was not so bad.

By October 1993, I personally received a very alarming call from the British diplomats advising me to change the hotel, and the chauffeur-driven car, and to avoid exchanging sensitive data during telephone calls with the headquarters in Toulouse. I was informed that the whole team was seriously taped, with ears-dropping everywhere. There was a big difference with past experiences. This was not done by the local authorities.

We moved to another hotel, changed cars, avoided leaving any sensitive document unattended in the rooms, we limited the number of telephone calls, and we consulted with our top management through direct contact (each week, one member of the team travelled to Toulouse for 48 hours, with all the questions and came back with the answers). This situation lasted for close to three months.

We were fully aware that there was a strong US government lobbying behind Boeing, further enhanced by the huge debt the kingdom had contracted vis-à-vis the US through the purchase of several military hardware, and particularly of a huge batch of F16 fighter aircraft. Some suggested that the amount of the debt was in the range of 6 to 8 billion US dollars.

Around Christmas 1993, we came back home for a break. Before leaving, I had a meeting with one of the top civil aviation officials to discuss some regulatory aspects of the setting up of an Airbus-authorised Training Centre (Saudia follows the US FAA rules and not those of the European civil aviation authorities).

During the discussion, he asked me about what would be needed to be done to cover different types of Airbus aircraft, mentioning precisely the A320 and the A340. I showed a surprise, telling him that it was unlikely that we win the competition.

He smiled and said to me:

"If you believe in Allah, trust Him; He could bring you good news. You never know."

I was amused by this statement. I tried to cross-check with other sources, and the conclusion was that our chances were *not bad*.

I left it there. There was nothing else to do. Meanwhile, Jean Pierson confirmed to me that the French Prime Minister, Edouard Balladur, was supposed to visit the kingdom in early January. We put a lot of hope that this visit will bring a positive conclusion to all our efforts made during more than a year and a half.

The first diplomatic rumours were tinted with some optimism. The visit took place around 9 January 1994, but nothing happened during the 18 hours the prime minister spent in Saudi Arabia, neither for Airbus nor for the several French manufacturers concerned by some big military projects like Sawary2.

There was no proper debrief from the entourage of the prime minister.

A few weeks later, the world listened, with a huge surprise, to President Clinton announcing in the presence of Prince Bandar Bin Sultan, ambassador of Saudi Arabia to the US, the CEOs of Boeing and McDonnell Douglas, and several other US officials, that Saudia, the airline of Saudi Arabia, had decided to buy a mixture of US-made aircraft from both manufacturers, without any further indication than the types: B777 and MD90s and MD11 Freighters.

The respective numbers were not announced (a *premiere* in the commercial aviation world). The President of the airline, Captain Mattar, was present, but he was not the master of ceremony.

This was also the first time the President of a country announced the decision of a commercial entity of another country to buy goods from suppliers from his own country. This was a real innovation in terms of business dealings. The US government showed that it did not care about any international rule.

We were so puzzled by this unprecedented announcement that it took us some time to grasp what really had happened. We did not need to carry out our own investigation. The international media, from US TV channels (particularly

CBS, through their programme '60 Minutes') to major international newspapers and magazines, disclosed the full story.

I just relate to what these media reported. They explained how the NSA, through their intelligence system *Echelon*, intercepted the telephone conversation between Prime Minister Balladur, from his flying Falcon 50, and President François Mitterrand, how President Clinton was informed, and how he called the King of Saudi Arabia with a *bargain*: if Saudia goes for Airbus, the F16 debt amounting multi-billion US dollars will be demanded immediately.

Another element of explanation was the forthcoming merger between Boeing and McDonnell Douglas. It seemed that the Saudi deal was the facilitator of such a merger!

In addition, and according to the same media, President Clinton accused Airbus of trying to corrupt some officials. At the level of our team, we only knew about the need to have a mandatory and legal sponsor (no foreign company can do business in the country without having a Wakil or legal representative), as per local regulations. President Clinton seemed to know more than the employees of Airbus.

But one question is still haunting me: who were the several Saudis I met after the deals were signed with Boeing and Douglas, and who told me that they had been selected to supply services ranging from aircraft cabin outfitting design to spare acquisition and monitoring, to documentation printing, etc.? I thought all these services would have been included in the price of the aircraft.

Airbus was crucified for, to my knowledge, having an official sponsor (who was not a government official) and for getting the help of executives of one of its mother companies.

But what was the role of the people mentioned above, and why they were there?

We eventually understood what had happened, starting with the strange criticism about our handling of the sales campaign to the warning from the British diplomats in October 1993.

This deal was doomed for us from the beginning. We ran for the fun. For months, I had a paper clipped on my office door with the following inscription in red, *Clinton m'a tuer*, meaning *Clinton killed me*, using the same wrong spelling of the word tuer (supposed to be tué), like in the murder case of a lady in Nice, next to whom they found the inscription, "Omar m'a tuer, hereby blaming the gardener.

Needless to mention that I was receiving expressions of sympathy and solidarity for weeks, not to say months. It was like receiving condolences. We were used to lose sales campaigns; it is part of our job. But this one was very painful because it was unfair.

In any case, the US inaugurated new rules to do business. Fair competition became a simple joke: the rest of the world must abide by all the regulations, but the US can do whatever it likes (what is the real definition of political blackmailing?).

We stayed away from Saudia for close to 12 years. We came back, as of 2007, and participated in new calls for tender with a positive outcome. Finally, we succeeded to sell A320s and A330s in large numbers.

Here again, patience and adaptation to the ever-changing environment enabled us to make a comeback. Because in this kind of activity, nothing is ever lost forever. Even if you don't get the deal, the airline gets to know you, gets familiarised with your ways, and, sooner or later, also because they need the competition, they will open their doors again and give you an opportunity.

Always remember you may lose one day but win the next battle.

The early 90s saw also the beginning of a big reshuffle in the landscape of air transport in the Middle East and particularly in the Gulf. To give a simple description, we can say that before the mid-70s, the dominant airlines were Iran Air, MEA, Gulf Air, Kuwait Airways, Iraqi Airways, Saudia, and Egyptair (we here exclude the airlines of North Africa, which were more oriented towards Europe).

From the mid-70s and following the Lebanese Civil War and the Iranian revolution, the situation changed, and the key players became Royal Jordanian, Kuwait Airways, Iraqi Airways, Saudia, and Egyptair. In this respect, we have to underline those airports like Amman, Bahrain, Baghdad, and Kuwait became important hubs for flight connections between Europe, the Middle East, and Asia. They took a huge chunk of the traffic of Beirut and Tehran.

The arrival of Emirates in 1985 took, initially, a small part of the Gulf Air traffic, but globally the traffic grew at higher rates in the Middle East than in the rest of the world.

Many people were predicting either the collapse of Gulf Air, because it was a joint venture owned by four countries (Abu Dhabi, Bahrain, Qatar, and Oman), or the failure of the Emirates. Neither happened, and both airlines enjoyed healthy growth. The secret was the intelligent development of the Hub concept, combined with the offering of much better services than the European and African competitors.

From the early 90s, many airlines in the region renewed their fleets, improved their services, and expanded their networks. Others suffered from the consequences of wars and revolutions. (Iraqi Airways, Iran Air), and some recovered from years of wars and troubles (MEA, Kuwait Airways).

But the most striking fact was the creation of several new airlines by governments or by private investors. In Egypt, for example, we saw the birth of a dozen of private charter airlines. Nevertheless, the two main events happened in 1993 when the Gulf witnessed the birth of Qatar Airways and Oman Air.

The creation of these two companies was seen as a sign of the final collapse of Gulf Air and was not understood by many analysts who considered that there was a limited volume of traffic and that the increase in the number of airlines would only reduce the respective shares of business.

I remember our marketing department making an exhaustive study which showed that some of the Gulf cities, mainly in Oman and in the UAE, were not properly served by the existing airlines and that there was a big potential of transit traffic which was not satisfied.

That means that there was room for more capacity deployed in the region. This could explain the bullish attitude of the Dubai leadership, as previously described, and the decision of the Qatari and Omani authorities to create their own airline.

Despite the arrival of three competitors (Emirates, Qatar Airways, and Oman Air) within eight years, Gulf Air continued to operate and progress, ultimately under the sole ownership of the state of Bahrain. But the most visible element of the multiplication of the number of airlines in the Gulf was that the traffic increased beyond any forecast or expectation.

The magic formula, combining modern aircraft with convenient connections, high-quality service, both on the ground and on board, and the utilisation of modern and comfortable airports, created a strong appeal to passengers around the world to use the Gulf carriers.

Today, we can say that the airlines of the Gulf, following the lead of Emirates, have created a sub-league in the airline industry, which could be defined by a totally new concept of business. Within less than 40 years, these airlines have done for the market more than what the European and US airlines have done for more than a hundred years.

Somehow, the air transport industry has seen two major trends developing since the 80s: on the one hand, the deregulation and the creation of the so-called *low-cost* airlines (as explained later), and on the other hand, the top-quality airlines. Between these two trends, many legacy airlines in Europe and the US are still trying to find their way.

In my opinion, it is worth giving more details about the creation of Oman Air and Qatar Airways.

Oman had already an embryo of an airline with Oman Aviation, which was operating some turbo-prop aircraft to serve the oil fields and some remote cities. The country is quite big, and the distances are long. A domestic network was badly needed, and Gulf Air was operating only in very few cities.

Therefore, one can say that the decision was based mainly on economic and social reasons (the populations in the remote regions must feel that the state is present and that they are part of a unified country).

Of course, Oman is also a very attractive touristic destination, and there is a need to transport more tourists from all over the world within the country. Unfortunately, Gulf Air was considered not fulfilling this role in a satisfactory manner.

The decision of Oman was expected, and it was made in a consensual manner since the Omani government kept its share in Gulf Air up to 2007, the latter continuing to operate the long-haul flights to and from Muscat, the capital of Oman, until 2009.

For Qatar Airways, the decision was based much more on political considerations. The state of Qatar had a plan to develop a QATAR brand; therefore it needed some strong symbols. A national airline was one of them.

The country did not need a domestic network, but the authorities saw that Dubai was boosting its economy thanks to Emirates. They also noticed that the sponsorship by Emirates of several international events was a clever way to put the name of Dubai on the world map.

This was a constant concern for the Qataris:

"We must exist in the middle of a hostile environment. Our country must be continuously on the world map, and we must send reminders every day."

Therefore, they thought that they could do the same thing. I am convinced that initially, they were not too much concerned about the immediate profitability of the airline (it was nice to have).

This was the first version of Qatar Airways, under the leadership of one of the members of the ruling family. The airline leased used B747s and B727s and did not seem to have a proper commercial strategy. It started operations in 1994.

This experience lasted for close to three years, up to 1997, when the country leadership stepped in, appointing Akbar Al Baker as president of the airline and decided to buy new modern aircraft and put in place a proper aggressive commercial strategy.

I knew Akbar Al Baker since the early 90s when he was working for the Qatari Civil Aviation Authority. He used to attend some of the Gulf Air board meetings as a member of the Qatari delegation and to visit some airshows with the Gulf Air management. He was a young, ambitious man, and we developed a friendly relationship, which lasted for years.

During an informal lunch in a pizzeria at the Doha Sheraton Hotel, one week after his appointment, we discussed the Qatar Airways plans. I remember there were so many details I was not expecting at that early stage that I was obliged to take some notes on the paper napkin of the table. At the end of the lunch, I cut the portion of the napkin, which was full of my notes, to take it with me, under the laughs of Akbar.

Qatar Airways became a big Airbus customer with successive big orders, including the famous one for 80 A350s, signed in June 2007 at Le Bourget Air show, in Paris. It also signed many big orders with Boeing.

I personally witnessed all the first steps of Akbar in Qatar Airways, from defining the fleet plan to choosing the uniforms of the cabin attendants and fine-tuning the logo of the company. The A350 contract was the source of my first disagreement with him, and it led to the progressive weakening of our personal relationship.

After a few years, he arrived at the point where he asked to deal directly with the CEO of the company and even the CEO of our parent company. I lost contact with him from 2014.

I regret this evolution because I thought I had developed a close relationship with him. In any case, I respect his achievements in building a top-quality airline and wish him good luck in his future endeavours.

Overall, this shows how important it is to have a good assessment of the potential development of the market in a given region, a good understanding of the economic and political situations of the respective countries, and a very good *flair*.

This also means a good fit with the persons in charge and their confidence. In fact, it is each time a bet, but in order to make the right assessment, it is important to also fully understand the culture of the country and the region.

Another example of what can happen when trying to sell aircraft is what happened to me in Yemen. 1994 was also the year of the Yemeni civil war, the start of which I followed directly from the French embassy in Sanaa. I was visiting Yemen in mid-February 1994. I had several meetings with the airline and the government officials to explain some financing issues and suggest different solutions.

We were a team of four people. All the important meetings were finished on 19 February, and the three other persons flew back to Europe. I stayed alone for one more day to have a debrief with the ambassadors. We badly needed the European Credit Agencies' support, and for this, the role of the ambassadors can be important through their reports. On the 19th evening, I had dinner with Mr Laugel, the French Ambassador, in his residence.

I had already seen two ambassadors during the day, and my tour of the ambassadors would finish with this dinner. Therefore, I had a flight planned for the next day, but as a precaution, I had also booked a seat on the KLM flight leaving at 2 am on the 20th. At around 9 pm, in the middle of the dinner, Mr Laugel had phone calls with both Ali Salem Al Beidh, the Leader of South Yemen, and Ali Abdallah Saleh, the President of the United Yemen.

Following the two conversations, he came back very upset, saying with great anger:

"Al Beidh does not want to understand the reality on the ground and believes that his posture is strong enough to stop the war. The other guy is as stubborn as him and is adamant about keeping the status quo and the union as it is. The war is imminent. If I have some advice for you, leave the country as soon as possible. The tanks and the armoured vehicles have already started moving."

We finished the dinner in a hasty manner. He asked his driver to take me to the hotel to collect my luggage and bring me back to the embassy. I did all this in less than an hour. The city was empty, besides a lot of military and security vehicles driving along the main axis. Mr Laugel was fluent in classic Arabic and in Yemeni dialect. I was impressed by his ability to speak fast in Yemeni.

I was admiring his desire to broker an agreement between the two sides to avoid a stupid war. As he expected, the two sides did not change their mind. At 11:30 pm, his driver took me to the airport. One guy from the embassy came with me to help me check in and go through the immigration and security checks as fast as possible.

Just around 1 am, I was seated in the lounge waiting for the boarding. The KLM B767 was there, and the airport was operating normally, apart from an increasing number of soldiers taking positions in different locations inside and outside the building and on the tarmac. I saw some of them around our aircraft.

When I landed in Amsterdam, I saw on CNN Breaking News that some fighting had started in Yemen. Here is another country where we had to put our activities on hold.

I saw Mr Laugel some years later and he was kind enough to explain to me the full picture. He was really a dynamic ambassador, completely integrated into the country of his assignment.

Yemen was the most exciting experience we had, at the Middle East business unit, to sell aircraft. We started dealing with two different countries: North Yemen with its capital Sanaa and South Yemen with its capital Aden. The first was very conservative with a basic liberal economy and a Middle Eastern-style society, and the second being influenced by Communism with a socialist economy and a Western-style society. Each one had its own airline.

In May 1990, the two countries reunited but kept two separate airlines. Despite the reunification, many observers were doubtful about the sincere willingness of the two leaders to cooperate and work together for the benefit of the country.

There were constant disagreements between the two and a serious clash was expected at any time. Nevertheless, during the four years between the reunification and the civil war, we continued to work with the two airlines. I personally travelled to Aden and to Sanaa several times.

To avoid any conflict of interest and any blame from the two airlines Alyemda and Yemenia, we decided to continue to have two different salesmen: Harry Kornberg for Yemenia and Stuart Wheeler for Alyemda. Harry, whose exact full name is Enrique Kornberg Gomes, is a Spanish citizen and Stuart is a British citizen.

Harry is a real adventurer who does not hesitate to take a few days off to climb a mountain in a newly visited country or to sail around the world with his whole family for close to a year. I had several oppotunities to experience Harry's adventurous spirit, and I don't regret it because I always learnt something while sharing his adventures.

But for once, I missed the most dangerous and breathtaking adventure Harry and Stuart lived in the middle of the Yemeni mountains. For some reasons, I could not recall, both decided to travel by car, alone, from Sanaa to Aden. During these days, the mountains were very dangerous because of the presence of fighters from different tribes, mainly in the north part of the country.

During their long drive, Harry and Stuart were stopped by armed men and questioned about their nationalities. Harry, who is Spanish, looks like an Arab and understands and speaks some Moroccan Arabic, thanks to the 20 years he spent with his parents in Morocco, was considered a cousin, but Stuart was considered an enemy.

One of the tribesmen asked Harry: "How would you like me to kill the Englishman?"

"There are different options, I shoot him with my Kalashnikov, I slash his throat, I dig a knife in his belly, or I can even kill him with my hands."

Both were terrified, but Harry took advantage of the capital of friendship he seemed to have with the group and dared to question their hospitality spirit.

After a long debate, during which Stuart was frozen by the fear of being executed in this remote area for stupid reasons (he must have regretted accepting Harry's suggestion to discover the beautiful scenery of Arabia Felix), the tribesman decided to not harm Stuart and they invited Harry and Stuart for a Yemeni meal.

The duo lived the scenario of kidnapping tourism, which was flourishing for a while in Yemen. Both kept a good souvenir despite the stress and the fears. Again, this is what one can be exposed to when selling airliners! Not something you would have expected when signing up for the job!

I lived a much more peaceful experience in a Houthi (or Zaydi) village, where I was invited by my driver Hassan, who used to give me a ride each time I visited Yemen. His village was not very far from Sanaa and I spent a night there. I enjoyed the fantastic hospitality, and interesting religion, history, and poetry discussions.

I was shocked by how teenagers were able to master Kalashnikovs like experienced warriors, with high scores in shooting on fixed and mobile targets. I had no clue why they had this culture of playing with deadly weapons.

I knew that Yemenis always have guns with them, like cowboys in Western movies. Hassan used to have one Kalashnikov inside the car and another in the car trunk. The recent war in Yemen made me understand the passion of Yemenis for guns.

I also lived a funny event with Jean Pierson during our joint and only visit to Sanaa. I had a very serious cold and I could not cure it. While we were visiting the old city of Sanaa, we discovered an old oil mill operated by a camel, housed in a sort of cave with a very small door.

Jean was the first to introduce his head in the small door and he got out quickly, but he grabbed my neck and pushed my head into the door telling me:

"I found the best remedy for your cold."

Despite the nose blockage, I could smell a strong odour of dejection and urine. Jean was right; this was better than any medical spray. Both of us felt sorry for the poor camel and admired the ingenuity of the owner of the mill who could get a fully functioning mill from a very limited space. Poverty generates creativity.

During the flight back to Toulouse, Jean told me to look for alternate solutions to satisfy the Yemenia requirements for aircraft because the country was poor with limited financial resources.

He added:

"I like to sell, but we must take into consideration the realities of the country."

This illustrates once more the need to fully understand the countries' environment and consequently, be very creative in your approach.

We cannot leave Yemen without mentioning the practice of chewing Qat. Every afternoon, Yemenis gather to share chewing the leaves of Qat, with some drinking alcohol to enhance its effect. I had an opportunity to sit in a *majlis* (gathering) of Qat chewing. I could barely and partially chew two leaves before spitting the content discretely and asking for a Coca-Cola to diffuse the taste.

I could notice that the people around me were enjoying it and became more inspired to recite poems or even to establish a dialogue with poems. The Qat chewing was a democratic exercise. I saw government officials, businessmen, teachers, and farmers sitting together around a bunch of Qat leaves and enjoying chewing together.

I could not refrain from laughing when I saw all these men's swollen cheeks moving in harmony, full of Qat leaves. If I must keep one single funny image of Yemen, I'll keep the image of a security guard, in front of the house of a VIP, with a Kalashnikov in his hands and his cheeks full of Qat.

As is customary amongst manufacturers and suppliers, as well as for airlines, as part of the overall business relations and to allow some unrestricted, relaxed, and frank exchanges, special events are often being organised, involving airlines, aircraft manufacturers, and suppliers. This was a common practice at the time, while things are much more restricted nowadays.

One of the most famous ones is the yearly *Conquistadores del Cielo,* which was initiated by the American manufacturers, and which involves said manufacturers as well as their airline customers. And there are many more such events that take place on a regular basis.

After all, even top senior executives are human beings, and good relations amongst the leaders of the business community can largely facilitate mutual understanding and get things moving much more easily than many hours sitting in a board room. At Airbus, we also had such working gatherings, including with the leading engine suppliers, amongst which General Electric (GE).

In this context, since Jean Pierson took the helm of Airbus in 1985, he had agreed with his friend and counterpart in GE, John Rowe, to organise an Airbus/GE summit every year, once hosted by Airbus and once hosted by GE. The host company choses the location and takes care of the organisation and the logistics.

It was a challenging task to impress the American guests with new activities. When they are the host, GE often chose locations in the Americas, with activities based on cowboys or rancheros lives, on golf, on sailing, and on trekking, and the food is based, mainly on Tex-Mex.

That year, it was the turn of Airbus to organise the next Airbus/GE top management yearly gathering. Due to our common link with Tunisia, Jean Pierson asked me to organise it in Sousse and Hammamet in Tunisia.

With the organisation team, we had to be innovative and create activities unknown to the Americans. For the food, all was new thanks to the Tunisian cuisine, which was not known in the US. We organised camel rides, pétanque contests, lessons of Tunisian bagpipe (which is the national music instrument, going back to the Carthaginian period and which is similar to the Scottish one), and of course, the usual golf and sailing competitions.

The big fun was with the camel ride and with the pétanque. Some of our American friends who accepted the challenge of riding camels on the beach could not finish the set distance and elected to go for a swim. At the end, there were more ladies riding camels than men (at the time, it was also common practice to invite the wives—or husbands, because in the end, they suffer a lot from the stress their spouses live through in the office and from their frequent absences due to travel).

For the pétanque, Jean Pierson, who was a good player, was happy to defeat the people against whom he could not compete in golf, a game he considered only suitable for retired persons. He never deviated from this idea (and he would not play golf even when retired).

This yearly get-together between Airbus and GE was a fantastic opportunity for the two leaderships to communicate in a relaxed atmosphere and in full transparency. I never understood why Jean's successors stopped this routine. To be fair, there are still regular meetings with the different engine manufacturers, but not in the format and duration of the ones that used to be organised with GE.

My perception is that the relationship with the suppliers, and especially GE, was not the same any longer. It shows how personal relations and mutual understanding, as established during such gatherings, can have a positive effect on business relations.

During this first period of Airbus, as much as we enjoyed the flexibility of the G.I.E., we suffered from the lack of strong support from the partner countries' governments. They tried to be supportive, but if they had to choose between supporting one national company or supporting Airbus, the choice was very simple and clear.

It was better to have 100% of something than a chunk of Airbus. In a certain way, it strengthened our independence and our self-reliance. We were a truly European company before a strong Europe was born.

On the personal side, in 1994, I was head-hunted by Matra, the French defence company, led by the late Jean Luc Lagardere, which was known for its missiles. At the beginning, I went through a lengthy process of interviews by an external recruitment firm, before knowing the name of the company and being allowed to meet its management.

I was reluctant to leave Airbus because I got the feeling that I would betray Jean Pierson. But I was tempted to see how head-hunting works.

As soon as I got the name, I knew that I would not accept the job. I could not work for a supplier of death machines. But I decided to go through the remaining interviews with the senior management of the company. This could help me in the future, in case I decide to leave Airbus one day.

I had interviews with Noël Forgeard, Philippe D'Allest, Jean Paul Gut (whom I had met twice before, in Kuwait and in the UAE, because we were in the same hotel), and their HR Vice President.

The interviews went well. I had a mixture of curiosity and reluctance, but I was very impressed by Mr D'Allest, who did a lot for the French and European space programme, and I had with him a very interesting discussion, like the one I used to have with my professor, Mr Thourel.

I discovered Noël Forgeard and knew what sort of manager he was from his comments, body language, and references. The discussion with J. P. Gut was very friendly, and I was expecting all his questions.

Finally, they made me an offer with a slightly higher salary than what I had with Airbus and gave me a few days to reflect on it and give my final answer. I knew what to answer but preferred to wait for a few days. While I was preparing my answer, I met a friend who had a lot of interactions with Matra through his responsibilities in the defence industry.

I asked for his advice and his answer was very short and very simple:

"It is not a job for you."

I wrote my refusal letter and stayed in Airbus, but my road crossed the road of Forgeard, Gut and part of the Matra management a few years later, and it was not my choice.

1994 was a very tough year, during which I lost close friends from both the flight test and marketing departments, following the crash of an A330 during a flight test at Toulouse airport, on 30 June 1994. I was in my office when I heard a loud explosion at a close distance.

It took me a few minutes to understand the cause of the explosion, thanks to the immediate information given to me by one colleague. That day, I was expecting a delegation from Gulf Air in Toulouse to prepare a delivery of an A340 during July.

In this crash, I lost my friend Nick Warner, chief test pilot, and Jean Pierre Petit, test flight engineer. I knew both very well, because I did several demonstration flights with them, either together or separately. I really enjoyed their company and appreciated their professional and human qualities.

Nick, in particular, taught me a lot about the certification process of several famous British aircraft, due to his previous experience as Civil Aviation Authority (CAA) chief test pilot in the UK.

The flight test and flight training department paid a heavy price for the company. The co-pilot of the A330, Michel Caïs, who had been seconded by Air France, some pilots from Al Italia, also died in that crash, as well as two members of the marketing team. But they were not the only ones to lose their lives in such a way. Other pilots lost their lives on test flights or in other missions abroad, such as investigating airline crashes.

This was the case for one pilot, Captain Iraj Fatimi, who died after a heart attack on the site of an aircraft crash in Hungary. Iraj could have been the hero of an adventure movie about the Iranian revolution. He was the pilot of the Shah of Iran, and he had to leave his country in very complicated conditions.

He escaped shortly after the precipitated departure of Captain Ziegler from Tehran, the next day after the fall of the Shah. Captain Ziegler was the last passenger to board a fully booked Swissair flight, leaving Tehran to Zurich.

His arrival in an Iran Air ramp car at the bottom of the stairs and his fast climbing of the stairs could have been sequences of the movie *ARGO*. Captain Ziegler was supervising some line training of Iran Air pilots, as part of the A300-B4 purchase by Iran Air.

For Iraj, it was more complicated. He had to hide and leave the country through remote areas, driving carts, riding donkeys, and walking. It was a real ordeal. He was among the wanted people by the new revolutionary guards. He could risk his life at any time.

Finally, he succeeded to get out of Iran and settling in Europe, where he rebuilt a new life. He ended up joining the Airbus Flight Department, where he held several key positions. One of his main concerns was his inability to go back to Iran.

Luckily, the situation evolved, and he could finally return to his country to train Iran Air pilots. The day he got his authorisation, he was happy like a kid in front of a new toy. Iraj was a real added value to Airbus and contributed a lot to change the perception that only *white* can fly.

In this respect, I can only pay tribute to Jean Pierson, who did not hesitate to recruit a respectable number of North African and Middle Eastern managers and pilots.

Following me, we had Mohamed El Borai, an Egyptian, who became later Vice President of Customer Support; Noredine Ouabdesselam, an Algerian, who was Sales Financing Director, Captain Belguedj, an Algerian, chief pilot instructor; Captain Aws Al Gemlas, chief pilot instructor; Abdellah Sbai, who was my deputy; and many more.

The number of Airbus employees from the MENA region grew tremendously over the years, to the point there are, now, a few hundred of them working in France, Germany, Spain, the UK, and other locations around the world, working for the whole Airbus Group. Their presence certainly contributed a lot to a better understanding of that region and to Airbus gaining an ever-increasing market share there.

During the first half of 1993, and while we were in the middle of the Saudia campaign, I used to frequently see Jean Pierson, to coordinate with him. Stuart Iddles was, of course, involved as Head of Commercial, but I noticed that Jean preferred to deal with me directly, although Stuart was directly involved. Was it because of my origins?

Progressively, I sensed a serious cooling in his relationship with Stuart. On a few occasions, Jean was critical of Stuart's decisions or analysis. I questioned Stuart, and he acknowledged that things were not going well with Jean. To help mend fences between the two, I organised a relaxed long lunch between them with their spouses, in my house.

During the month of July, before the summer holidays, at the end of the lunch, the two went out in the garden, smoking respectively Gitane cigarettes for Jean and Cohiba cigars for Stuart. They were laughing, and I thought the crisis was over. Apparently, the reasons behind the deterioration of the relations between the two were more serious, and my initiative had a very short life.

By the last quarter of 1993, we had many challenging campaigns, and Jean was in the middle of negotiations with the partners for some cost reductions and several other things. Stuart Iddles, the Head of Commercial, was mainly focusing his attention on the Saudia campaign and a few others in Latin America. Since BAe was our main support in Saudi Arabia, Stuart was the coordinator between the two companies.

During a very difficult meeting about Saudia, Jean exploded about our inability to find additional early slots for the airline, because they were booked for other airlines, and said:

"This commercial directorate needs a complete reorganisation."

Everybody around the table was shocked, and Stuart became blemish. That day, I knew that the game was over for Stuart.

Stuart was a kind, friendly person, a boss close to his team. He travelled a lot in his life, and he spent a lot of time abroad, including a few years in Tunisia. He was fluent in French and knew some Tunisian. Since his wife was Mexican, he also spoke Spanish. He was perfectly multilingual and fully aware of the problems of the developing countries.

We built very close relations, and we enjoyed visiting each other and playing golf together. Despite differences in the management styles, Jean Pierson and Stuart Iddles were both very human, friendly, open to the world, and enjoyed good food, and in particular Tunisian food. Stuart has really succeeded in creating a warm and familial ambience between all the members of his team.

Jean Pierson never explained the reasons behind the departure of Stuart Iddles, but I never heard him criticising him.

In the end, Stuart left just after the Clinton announcement about the Saudia order. Sir Charles Masefield from BAe was appointed as new SVP Commercial, as of 1 March 1994. This confirmed that the commercial directorate was the turf of the British. Sir Charles rented a house not so far from mine, and I had the opportunity to know him and to have multiple discussions with him.

He started his new assignment while being well perceived by the heads of world regions and the commercial directorate employees. But his stay in Toulouse was short. By the summer of 1994, he was offered the position of head of UK Defence and Security Exports (UKDSE), known also as Defence Export Services Organisation (DESO). He accepted the position and was subsequently knighted. He left Airbus at the end of August 1994.

Jean replaced him by John Leahy, a US citizen who was president of Airbus Industries North America (or AINA), as of 1 September 1994. As Jean explained, after BAe had proposed and seconded two senior executives to Airbus for the position of Head of Commercial, and that they had "kind of failed," he felt now free to elect—and propose to the partners—the person from within Airbus whom he thought was most fit for the job.

And in this respect, Jean was once more right: John ended up being the longest-serving Head of Commercial in the industry, holding the position for over 20 years and through quite a few managerial changes.

John had joined Airbus Industries North America (AINA) in 1985, a few months before I had joined Airbus in Toulouse. Jean used to oppose us against each other by comparing our sales performance in the US and in the Middle East.

We were the two first to have mobile phones (it has to be said that Jean initially refused to have a mobile phone, as he did not want to be bothered by unexpected calls), and Jean's best game was to compare our telephone bills and make ratios with the volume of sales.

Despite working in different regions of the world, we had developed friendly relations, and I went to visit John in his office in AINA at least twice. John worked for Piper Aviation before joining Airbus and is a professional pilot.

I was pleased with this choice and did my best to support John in his new job. Despite the friendly opposition Jean had created between the two of us, I had no competitive spirit vis-à-vis John and I respected his leadership. Nevertheless, I had to educate him on the political, religious, cultural, social, and human specificities of the MENA region.

To be fair with John, he gave me total freedom to run the show for some years, before starting to get more involved, following the change of management in 1998 and the subsequent change of ownership of Airbus. This did not prevent us of remaining good friends until the end, and I fully respect and admire his achievements.

Among the big changes brought in by John, we can mention the increased role of communication and the more professional setup of the yearly commercial symposium.

Bob Alizart, who was the SVP Communication, and Barbara Kracht, the VP Media and Press Relations, were doing a good job, but Jean Pierson and Stuart Iddles were not too much present in the media, and this was on purpose. When John arrived, things changed, and he became a regular feature in the headline news.

This trend was further accentuated, first with the arrival of Noël Forgeard in 1998 and, second, with the transformation of Airbus into an integrated company.

As of the mid-90s, the territory I was overseeing was being progressively extended, to include some Asian countries like Pakistan, Bangladesh, Malaysia, and Indonesia. This gave me the opportunity to travel often to these countries. I discovered new people and new civilisations.

The fact that these countries had Muslim majorities facilitated a lot my access to them. First, they use a lot of Arabic words, some of them written with Arabic alphabet, and some social expressions are absolutely identical.

One of the major developments, in terms of sales, was the opening we had in Syria. For several political reasons, Airbus was allowed to sell aircraft to this country after a period during which the situation could be described as "grey."

There was the intervention of Syria in the Lebanese civil war, which caused a lot of Western negative reactions against the Syrian regime of Hafez Al Assad, the father of the Bashar Al Assad who was recently overthrown by a popular uprising. But there was also the participation of the Syrian army in the coalition to liberate Kuwait, which brought some positive spin to the Syrian-West relations.

Therefore, we could start a sales campaign with Syrian Arab Airlines or Syrian Air. This campaign was very difficult and lasted for more than two years. The country was closed to business dealing with the West for a long time, and we had to go through some learning process.

It was shocking for some members of the management who were present, during the purchase of the B727, in the early 70s, at prices around 6/7 million US dollars, to see a very much higher price for an A320, for the same number of passengers to be transported. We had to go through detailed presentations covering technical, operational, commercial, and financial aspects, to explain what the changes were and to try to convince them.

Many of the airline managers were highly educated and were used to dealing with the West, but the decision process was seriously tinted by the Soviet influence they had experienced for decades. On the technical side, we found very competent and knowledgeable personnel in maintenance and flight operations.

This long and exhausting campaign allowed us to have a closer view of the way the country was ruled. Everything had to go to the top for approval. Even the prime minister's decisions could be questioned.

We spent hours and days waiting for meetings with ministers or the prime minister, and when we got them, we had, very often, the frustrating sentences:

"We will check and let you know. You have to come back."

The airline had no power to decide anything.

Finally, and after months of discussions and negotiations, a contract for the sale of eight A320s was signed. It was de facto, a no-choice for the airline to buy Airbus, but the contract was negotiated as if there were 10 competitors. This was rewarding, despite the difficulties. It is always better to win a deal after a serious fight. The bargaining in the Middle East is famous and deserves a dedicated book. You can never say it is finished.

The opening to Asia started with a working lunch with Lee Kwan Yew, the ex-prime minister of Singapore. For some reasons, I was in Singapore for the Asian Aerospace Airshow (the previous version of the current Singapore Airshow, as Asian Aerospace was moved to Hong Kong as of 2008) at the same time as Jean Pierson.

Airbus had helped the government of Singapore in strengthening its aerospace industry, and Pierson was a much-welcomed person in Singapore. He was invited to a lunch with the ex-prime minister and was allowed to have two or three people with him. He kindly invited me to join him.

The lunch, which was in reality a working lunch involving many foreigners, with a presentation from Mr Lee Kwan Yew, followed by Q&A, was a real opportunity to debate with the founding father of Singapore about how he and his team built such a fantastic success. I believe I attended the most enlightening conference on how a clear vision and a strict road map can achieve miracles.

We had a detailed description of the process of the creation of the state of Singapore, from the first step for independence in 1959, to the split from the federation of Malaysia in August 1965.

He explained the vision he set for his country (I understood that they set targets in the frame of what was called Vision Singapore 2000 or simply Singapore 2000), the harsh actions he had to undertake immediately, and the clear and strict road map he set for the decades to come. He explained the difficulties and the challenges the country faced.

A very interesting debate followed on the compatibility between full democracy and quick development. I have to confess that I was impressed by the explanations given by Mr Lee Kwan Yew about the necessity to define priorities and to go step by step, while ensuring justice and equal chances for all the citizens.

He agreed on the need for democracy to ensure the sustainability of the state. But democracy must become a natural state of mind and behaviour for each citizen.

I still remember this debate, which allowed me to learn more about Singapore and the reasons behind its successes. I noted that the authorities of Singapore made a strategic choice to be the Asian hub, and they bet on services and high technology.

They considered aviation as one of the principal tools to reach their objectives and made it a priority for the country. Hence, the world status of Singapore Airlines and Changi Airport, which are the results of strategic choices.

From this first active and enriching part of my career in Airbus, I learnt many lessons. One of them was linked to the fears I expressed when I graduated and when I was doubtful about being able to take responsibilities based just on the diploma. In many countries, the level of education is considered a sufficient criterion to give the helm of big companies and organisations to people who have no capacity for leadership and are unable to manage a crisis and/or a project.

If this trend could be understood in developing countries, in their early days of independence, when highly educated people are scarce, it cannot be accepted in developed countries. Unfortunately, many airlines and industrial companies in aviation were put in the hands of supposedly highly educated people, who simply failed in managing them and, even worst, caused their collapse.

This was mainly true for government-owned entities. I have dozens of examples in mind, and I still question myself how this could happen, especially in European countries. This comforted my belief that a diploma is a simple paper, with a lot more being required to make somebody a good manager and ultimately a true leader. That is why there are a lot of CEOs, but only few leaders.

In this instance, Jean Pierson, is a typical example. He did not have a diploma from one of the French top elite schools such as ENA (École Nationale de l'Administration) or *X* (Polytechnique), which provide most of the French senior executives (usually also government appointees), but he definitely was a true leader.

Unlike those, he was neither arrogant nor did he talk down to simple employees. He always remained easy-going and easily accessible and enjoyed simple things of life, like eating couscous in a small bistro with some colleagues. Through this unusual attitude, plus his personal hands-on knowledge of what goes on in a factory, Jean inspired and motivated the Airbus teams much more, in my humble opinion, than many highly graded ones.

Airbus Corporate Jet (ACJ) (Magic Carpet Ride—Steppenwolf)

Both Airbus and Boeing had experience in selling their products for private users and to heads of state, to be outfitted with special cabins. Airbus had already sold some A300-600s, A310s, A320, and A340s for such use. This activity allowed us to have access to several head of states, senior princes and prominent businessmen.

In 1996, Boeing decided to launch a dedicated private jet based on the B737-700, called Boeing Business Jet or BBJ. Very quickly, the sales took off, and it appeared that there was a good market for this size of aircraft, beside the other business jets like the Gulfstream, the Bombardier Challenger, the Dassault Falcons, the Lear Jets, etc.

In the Middle East, we had the experience of such demand, and we were pushing the Airbus top management to do something with the A320 family. Despite the Boeing move, there was no reaction within Airbus. Noticing this lethargy, we decided, myself and my colleague Harry Kornberg, who was a sales director for Egypt and other countries and had the experience of selling an A340-200 for a VVIP, to promote the A319 as a competitor for the BBJ.

For that purpose, by early 1997, we made a preliminary study defining the product, estimating the market size, and proposing a name: Airbus Corporate Jet or ACJ. The study was presented to Jean Pierson and John Leahy, the SVP Commercial. Both liked the idea and instructed the marketing department to do a comprehensive study.

Meanwhile, we approached the engineering department to fine-tune the specifications, and we came up with the number of additional auxiliary fuel tanks, additional water quantity, etc., that would be needed.

In the second quarter of 1997, the ACJ was born, and we made our first sale to a Middle Eastern customer.

Subsequently, Jean Pierson and John Leahy decided to give me the responsibility of creating the ACJ sales business unit, under the umbrella of the Middle East team. The end result was that Harry became de facto the head of this small unit, in addition to his other responsibilities.

The activity took off, and we scored some high-profile sales, to the point that there was a decision to create a dedicated business unit.

This happened within a few months and coincided with other major changes. The only problem was that Jean Pierson decided to appoint another person in lieu of Harry, who, regrettably, left the company soon after.

When I was appointed Senior Vice President Sales (as explained later), I kept the supervision of the ACJ team, and this situation lasted up to the end of 2001. During this period, we achieved a very good level of sales of different types of aircraft for corporate and private use. They included in the end all the Airbus aircraft types, with a focus on A319/A320/A321, but also the A340 and others: ACJ A320, ACJ A340, and so on.

From 2002 to May 2008, I had other responsibilities, and the ACJ business was in the hands of the head of the business unit and its new boss.

In 2008, I recovered the ACJ business unit, and we extended its activities to customer support. The sales continued at an acceptable rate, and we ended up being nearly on par with Boeing.

In 2012, Tom Enders, the then Airbus CEO, responded positively to a study I had done in 2011, to create a full subsidiary of Airbus which would take care of all ACJ-related activities, from engineering to sales and marketing, up to cabin outfitting and customer support, with the objective to bring together all the existing resources which were scattered between different departments, and a pure industrial subsidiary in charge of outfitting.

Fabrice Brégier, who was the COO, played an important role in approving the studies and, later on, in materialising the project.

On 1 January 2012, I was appointed president of Airbus Corporate Jet on top of my position as President of Airbus Middle East (as will be explained later).

This dual responsibility lasted for two years, during which I was sharing my time between Dubai and Toulouse, and during which I organised this new subsidiary, streamlined the ACJ activities, and really made the world aware of Airbus's full commitment to the corporate and private market.

Together with my team, we by far exceeded the 100 sales and became a major player in the business jet market, despite the constantly growing size of the aircraft built by the traditional manufacturers. My agreement with Tom was to restructure and put on track the ACJ activity within two years, with the target to appoint a new president by the end 2013.

The plan was respected, and I gave the subsidiary to a successor appointed by the Airbus executive committee. Mission accomplished!

Among the major events I lived through with the corporate jet team, I must mention two as follows:

The Hainan Luxury Show, where we exhibited one of our ACJs and where I saw a luxury extravaganza. I never imagined that the citizens of a communist country would attend a show where displayed items are priced in millions of dollars and that they would not hesitate to wear expensive clothes and jewels and drive very expensive cars. I never saw such a concentration of diamonds in my life. Communism is shining.

Sanya, the capital of Hainan, also surprised me with the existence of a local Muslim community with its Halal restaurants and other community monuments where you could read panels written in Arabic, mainly on the road linking the airport to the city centre.

The visit to Sao Paulo, where we were supposed to participate in the local airshow by displaying one of our ACJs. Unfortunately, we could not land in Sao Paulo because of a long-lasting strike by the city's customs officers. For some strange reason, we had to pay a high amount of money, as a customs duty, to allow the aircraft to enter the country.

For this purpose, we had to fly down to the heart of the Amazon Forest, to Iguazu in the state of Parana, where the customs were not on strike and where we could obtain clearance for the aircraft. It was a very nice opportunity to visit the Iguazu Falls, which are one of the wonders of the world.

After this unforeseen visit to the heart of Brazil, we had a very quiet airshow. Some new taxation laws seemed to calm down all potential business jet buyers.

But the most curious comments I heard from a high-ranked officer in the army were:

"We are discussing with the army officers to avoid any social movements during the upcoming World Cup."

I thought soldiers never go on strike.

During this visit (which was not my first) to Brazil, I discovered some realities which explain the actual situation of the country.

The effect of the corporate jet activity on the rest of the company was never properly assessed. But a quick overview could show the following visible elements.

First: Selling a corporate jet can be an elegant way to approach directly the top decision makers within the authorities of the country, within big corporations, and within the community of high net worth individuals who, in many cases, own airlines. This proximity facilitated in a few cases the sale of aircraft to airlines owned or controlled by these individuals. It helps wipe out misperceptions and prejudices.

Second: Airbus became more focused on the cabin interior, and a lot of innovations could, as a result, be brought to passenger aircraft, improving tremendously the comfort and the level of service provided on board (individual suites, mood lighting, large TV screens, Wi-Fi on board).

In some cases, Airbus was the trendsetter. Aircraft cabin designers like Jacques Pierre Jean, who was the architect of several corporate jet cabins, contributed a lot in redefining the comfort of passenger aircraft, mainly in business and first classes.

Third: Some technical solutions developed for the corporate jets, like auxiliary fuel tanks, catering for bigger quantities of water on board, or installing showers on board, helped Airbus improving its offering to airlines.

On a more personal aspect, this long immersion in the world of the corporate jet allowed me to learn a lot about humanity and particularly about the people who rule the world, either politically or financially. One common comment is that all these people behave like children when it comes to choose and define the aircraft or the toy. There is a lot of excitement and curiosity.

But one can see the difference between those who have limited financial resources and those for whom the sky has no limits. The most modest in their demands are the corporations. They often look for an airline first-class comfort, with some space to meet and work. The billionaires like, very often, to replicate the comfort of their homes in the sky. The heads of state, on top of the comfort, have very stringent demands due to security considerations.

But there is one striking element. All these people give a very big importance to some small details in the aircraft, which we usually don't pay attention to. One day, the president of a country refused to accept the outfitting of his office on board because the seam of the edges of his chair was harsh.

He was right, and I was puzzled. How on earth, with all the tests, all the hands, and all the bottoms involved in the inspection of this chair, could no one have detected the problem, which was a real one? Despite their responsibilities and their busy agendas, these people can find time to look around and check on small details. It is something which attracted my attention. It shows that these people remain human, with all their qualities, but also with their defects.

Another aspect which attracted my attention during my years within ACJ was the somehow expected and strong interest shown by the luxury industry in our products (luxury attracts luxury). Some famous brands proposed to organise common events and/or develop joint products or services. Among those iconic products, we can mention the Richard Mille ACJ watch, which was the result of a cooperation agreement between ACJ and Richard Mille.

Similar discussions took place with a famous French luxury brand, but they did not materialise as expected. We can also mention the joint reception organised with the famous café-restaurant of the Champs Élysées, Le Fouquet's, in Sanya, the capital of Hainan Province in China, during the Hainan exhibition.

In a nutshell, the ACJ experience was very enriching and helped me learn a lot. It also helped Airbus enter a market which was previously dominated by US manufacturers, and to improve tremendously its offering in terms of passenger aircraft interiors.

Before closing the chapter on the Airbus G.I.E., it is worth telling the story of Pierson's trousers dropping, because it gives a very clear idea about his personality, his charisma, his stamina, and his ability to get out of the most complicated situations, something extremely important in this field.

Jean was heading a delegation to negotiate with Steve Wolf and his team a huge contract with US Airways. The negotiations were taking place at the airline headquarters, and Steve Wolf pushed hard to get additional concessions.

Suddenly, Jean Pierson stood up from his seat and said:

"Steve, do you want me to drop my trousers? I am already naked."

While he was saying this, his trousers did indeed drop. Everybody can imagine the hilarious reaction of the assistance. The whole assistance thought Jean did it on purpose to impress Steve. The reality was somewhat different. Following the recommendations of his doctor, Jean had gone through a drastic weight loss programme. Consequently, his waist perimeter had decreased tremendously, but he had kept the same trouser size.

Before joining the meeting, he went to the toilets and, apparently, did not fasten correctly his belt. Did he deliberately let his trousers drop, or was it an *accident*? No one will ever really know, as the versions Jean was telling varied depending on when and to whom he was telling the story. In any case, the trousers dropping made its effect, and the deal was concluded without further concessions.

This was Jean Pierson (RIP). He was the right person to put Airbus on track, by keeping the initial pioneering spirit while building the product line and expanding the customer base, to reach the critical size, and to eventually be on par with the number one in this industry sector, namely Boeing. He succeeded in his mission.

As already pointed out above, Jean Pierson was not an obvious candidate for the job at the time. He was neither an *X* or nor *Enarque* and, because of that, was not *proposed* for the Airbus Industrie top job by the French government, as was customary so far. Nor was he recruited externally for political reasons but proposed from internally within the GIE partners. This was, in a way, here too, the beauty of the GIE, by being able to get rid of political pressure!

After graduating from Supaéro, Jean actually had made his career starting at the Sud Aviation shop floor up into the ranks of senior management at Aerospatiale (into which Sud Aviation had been merged) and had therefore a lot of real hands-on experience. He was the charismatic and friendly leader who makes you feel confident and comfortable.

He could talk on equal terms with the *compagnons* (the name for the shop-floor technicians who work on the aircraft) or with any other colleague within the company, or at the other extreme, with a customer CEO or a President. But he was also a strong and decisive leader, who could explode with very colourful words and make you feel the fear deep in your bones.

He coached us, empowered us, and led us by example. I'll never forget the boss, the friend, the advisor, the companion for good dinners and luncheons, and the competitor in the famous *pétanque* game.

I'll never forget our joint visits to Tunisia and his pleasure to buy some local products. He was fond of Tunisian red Tuna, Harissa (Tunisian red spicy cream), and Boukha (the local liquor), to name just a few. He was as Tunisian as me, and this translated in our relations.

5.2 Airbus Corporation (More Professional, but Red Tapes) (Take a chance on me-ABBA)

In April 1998, Noël Forgeard took the helm of Airbus Industrie from Jean Pierson, who retired, after his third term, at the relatively young age of 58 years.

This was shortly after the buy-out of Mc Donnell Douglas by Boeing in 1997, which de facto created the duopoly between Airbus and Boeing. This competition between the two manufacturers, which was—and still is—fierce, became the spearhead of the global competition between Europe and the USA.

Four years after my interview with him, in the Matra headquarters, rue de Presbourg, near the Champs Élysées, in Paris, I am meeting him again, but this time we were supposed to work together for some time.

Noël brought with him some people from Matra, whom we called *the Matra or Lagardère boys*. Their number was limited, but they were in key positions and very close to the boss. Very quickly, I established relatively friendly relations with Noël, and I continued to work as usual, within the structure of the G.I.E.

Thanks to our excellent results in the extended Middle East region, and following a restructuring of the commercial directorate, I was promoted, in two steps between late 1998 and the summer of 1999, to the position of Senior Vice President of Sales, Business Development and Product Development in early 1999. I had responsibility for sales, worldwide except in the US.

This came after a few months during which I had been acting as deputy for John Leahy. My appointment took place at the same time as that of Christian Scherer as SVP of Contracts and Business Control. We formed a united pair and were happy to work closely together. I was his elder brother, and he was my younger sibling. We defended the same positions; we diverged only on very few occasions, on minor issues.

Even when, as it happened later, he was instructed to do things contradicting my decisions, he did not hesitate to tell me. He had, on several occasions, the courage to implement my decisions after our paths diverged. Christian remains a close friend, and I wish him all the success in his current job (he is now CEO of Airbus Commercial Aircraft).

On that occasion, I discovered the differences between the attitude of the four partners vis-à-vis a North African citizen, who was proposed for an executive position in a European company. Noël sent a fax, on a Wednesday morning at around 9 am, to the CEOs of the four partner companies and members of the Airbus Industrie Supervisory Board, to inform them and to get a sort of blessing from them for my appointment.

Noël got a very quick response from the British side, with a yes (I even got a phone call from the chief of staff of Bae President's office to congratulate me). The Germans sent their blessing two or three hours later, and the Spanish followed later in the day. In the two latter cases, I also got congratulations from a member of the respective cabinets. But, despite waiting up to the end of the week, nothing came from the French side.

It took, up to the next Wednesday, i.e., about a week later, and some reminding actions by Noël, to get a shy acknowledgement from the French partner. Following an investigation through some friends within Aerospatiale-Matra, I discovered that the top executives of the company had tried to find another option, preferably a French citizen.

Eventually, I got the job, but I did not realise that there were some people who would continue to try to put a spanner in the wheel.

One of the major consequences of the change of Airbus to a corporation was the substantial improvement in the salaries (I already mentioned that the salaries at the GIE were comparatively low, even that of Jean Pierson as CEO).

Following these new responsibilities, I embarked on a sort of world tour to visit the key customers. This was an opportunity to widen my horizon and discover new countries outside Europe, Africa, and the Middle East. Very quickly, I found myself involved in important sales campaigns, mainly in Russia, China, Hong Kong, Australia, Italy, and Brazil.

Russia was one of my first destinations after my appointment. I was excited to discover this big country and to renew contact with the land of my ancestors. My family traces back its origins to Crimea, then to Istanbul, before immigrating to Tunisia, by the end of the 18th century. We were Crimean when this country was occupied by the Ottomans before Tsarina Catherine of Russia invaded Crimea and annexed it in 1783.

Therefore, going to Russia was somehow going back to the roots of my family. I knew a lot about the Russian history and its traditions.

In particular, I knew that Russia is more Oriental than Occidental despite the fact that its population masters well a lot of elements of the West (music, arts, technology), and that Islam is a sort of native religion which existed in some regions of Russia for many centuries. In some places, Islam was even present before Christianity (as early as the 7th and the 8th century in Daghestan and other parts of the Caucasus). Therefore, Muslims are not facing the same negative reactions as in the West.

I adapted very quickly to the Russian atmosphere, helped a lot by my ability to read the Cyrillic alphabet thanks to the Russian language courses I had taken at Supaéro.

Despite my understanding of the Russian culture, I was however surprised and even sometimes shocked by some habits and behaviours inherited from the Soviet era. I was in Moscow less than eight years after the end of the Soviet regime, and people were in a difficult transition period.

Thanks to my relative understanding of the Russians, I succeeded to make many friends at different levels of the authorities and the airline management, as well as among other people I met in the hotels, bars, and theatres.

After my first visit and what I experienced there, I tried to convince my management that we must apprehend Russia and its people from a different angle, and not from a purely Western one, as I had figured out that some Russian attitudes are closer to those seen in the Middle East, in the Caucasus, and in Central Asia.

Unfortunately, this was not well understood, and I was blamed for letting my emotions take precedence over my reasoning. This sort of disagreement with my higher management (the CEO and the COO) about the realities of some countries became frequent as of 1998.

I do not pretend to say I was often right, but I discovered that there were some significant details which I could detect, thanks to the Oriental part of my brain and to my culture, which allowed me to have a clearer picture of the situation. In the case of Russia, the relations with our company experienced ups and downs, depending on who was running the show within Airbus.

The other country where I was happy to return to was Italy. A country I knew very well, which I had visited several times and which I loved so much. Thanks to the Italian TV channels which we could receive in Tunisia, most of my generation could understand and express itself correctly in Italian. There was no secret that I, very quickly, developed excellent relations with the management of Alitalia.

Here again, my management could not understand that the links between Tunisia and Italy, and between Tunisia and Sicily, were very strong (many common words, many common traditions, and a lot of food in common, mainly with Sicily).

I disagreed with the CEO on the strategy vis-à-vis Alitalia, and time eventually showed who was right.

During my tour of Europe, I discovered that some countries like Hungary and Poland were much closer to the US than to Europe. I understood the reasons, but I hoped that, progressively, these countries would become more European. One immediate consequence of this biased position was our inability to sell anything to these countries.

As much as Hungary evolved in terms of civil aircraft procurement and became a good Airbus customer, Poland, for its part, continued to show a certain reluctance towards Airbus. This is a clear illustration of how a strong diaspora can play a major role in shaping relations between the host country and the country of origin.

Even after being a member of the European Union, Poland remains very close to the USA and continues to buy most of its aviation hardware (both civil and military) from the US. I think it is a reality which is now accepted by the Airbus management.

After Europe, I started travelling to Asia, to visit the most important airlines of the region. One of the targets was Malaysian Airline System (MAS).

Being in London for the Farnborough Airshow in July 1998, I heard that His Excellency, Dr Mahatir Mohamed, the Prime Minister of Malaysia, was in town for the show. I, therefore, suggested to our CEO to try and meet him, in order to understand the expectations of the Malaysian authorities, their priorities, and the sort of cooperation they favour. Malaysian was partially private, but the final decision was in the hands of the government.

The airline CEO was one of the high-profile Bumiputera, or native Malay, who was at the head of a conglomerate owning several companies and having a sizeable share in the national airline. Since the 70s, the Malaysian government was implementing a very aggressive policy to reduce the gap between the Bumiputera community and the Chinese and Indian communities.

This policy was called the New Economic Policy (NEP), which was followed, in 1991, by the National Development Policy (NDP). It had several objectives: reduce poverty, break the work division between the communities, and harmonise relations between them while reengineering the economy.

This policy allowed many Bumiputera to become successful businessmen and participate in the economic development of their country, like their fellow citizens from the Indian and Chinese communities. For different reasons, Prime Minister Dr Mahatir was very often, personally, criticised for the NEP/NDP. This criticism was the cause of a strange alliance between him and me, during a formal event at the House of Lord, in London, a year later.

We succeeded in getting an appointment with Dr Mahatir for our CEO and two accompanying persons. I was one of them. We discovered a very simple man, very polite and soft speaking. Dr Mahatir gave us close to two hours for us to explain what we were proposing, what sort of cooperation could be established between Airbus and the country, and what were the financing options. He listened carefully and silently, asking some questions from time to time.

After we finished the meeting and were preparing to leave, he asked me to repeat my family name because he was not sure he heard it correctly (his economic advisor had introduced us to him earlier on).

I did, and he gave its exact translation in English to my CEO, adding:

"His real job should be in an Islamic university like Al Azhar."

Everybody laughed, and he asked me about my origins and background. This friendly attention surprised all the persons present, including the Malaysian side. We delayed our departure, and his protocol was nervous, but he did not seem to be concerned. This unexpected discussion lasted for close to 10 minutes.

Sometime later, I was in Toulouse, and I received a phone call from the Malaysian embassy in Paris, informing me that Dr Mahatir would be in Paris on a given date; and Her Excellency the ambassador (she was a lady) invited me for a dinner in honour of the PM.

Due to other very important commitments, I could not attend the dinner, but I could join Dr Mahatir after his visit to the Association of French Industry Leaders (MEDEF), and we had an opportunity to exchange on several subjects. He spoke to me about the new airport at Sepang and about several projects in the different provinces. He insisted on the need for Airbus to put some industrial facilities in the country, taking advantage of some of the existing factories.

He gave me a short lecture on the challenges he faced in ensuring a harmonious development for all the provinces.

When I told him that this was going to be more difficult than in Singapore, he answered:

"Please don't misunderstand me. I fully appreciate and value what has been done in Singapore, and I congratulate the people of Singapore and its leadership for their achievement. But the sizes of the countries and of the respective populations matter."

Thanks to this interesting meeting, we came to know each other much better. He was able to pronounce correctly my name in Arabic. I also discovered that he had some knowledge of Arabic, which he could read (it is worth mentioning that Malaysia uses both Latin and Jawi characters, which are derived from Arabic).

One strong sign of our growing relationship was demonstrated in Kuala Lumpur, a few months later, during a visit to Malaysia with the Airbus CEO and other executives.

It was the reaction of Dr Mahatir towards his assistant in charge of protocol, who introduced me as Mr Habib and added:

"Egyptian."

He corrected him on the spot, by saying:

"Mr Mohamed Habib Fekih, he is Tunisian."

Everybody was surprised by this intervention, and Noël Forgeard commented afterwards:

"Hey, I am impressed. He seems to know you very well."

After this visit to Kuala Lumpur, I was invited to a dinner at the House of Lords in London, in honour of Dr Mahatir. Many representatives of the UK business community were invited, as well as some government officials and some Lords. The dinner was preceded by an aperitif, and I had an opportunity to have a one-on-one chat with the PM.

Among the points discussed, he told me that, if Airbus agreed to sub-contract the manufacturing of some relevant A380 components to Malaysian companies, he would eye favourably the acquisition of such aircraft by the national airline, provided, of course, that an aircraft of such a size were to be needed. He added, with a smile, that he did not like to interfere in the airline business, otherwise he would get all the blame in case of a problem.

During our more than 20- minute chat, we spoke about the negative attitude the British media and political establishment seemed to have against his NDP, which promoted a lot of Bumiputera to become high-profile businessmen.

I questioned him about the attitude of the British, and he gave me simple and diplomatic answers, but he expressed a certain disappointment about what he called a misunderstanding and said:

"Maybe, I could give more details over dinner."

I told him, then, that I could even provoke him on the subject. He liked the idea and explained to me that there will be a speech from the host, then he will say few words and will conclude by asking if they were questions. At that moment, I could jump in and ask him about the NDP. We agreed on the terms of the question. I asked him, then, why he would like me to ask.

He answered that he does not like to get questions from those who criticise him before fully understanding the purposes of the NDP and asking for clarifications. It was fully understood, and I acted as agreed. I was the first to ask a question, and everybody around me was in a state of shock. How did I dare to be so blunt?

Dr Mahatir was a real maestro in his answer, which was very clear and exhaustive. He passed all the messages he wished to pass. At the end of the dinner, I had some congratulations from two Lords and the Speaker of the House of Lords himself, who said to me that I was courageous.

This trusting relationship between Dr Mahatir and myself was demonstrated again a few months later during the 1999 Langkawi Airshow when I was invited to make the formal tour of the exhibition as part of the official delegation. As such, I had a small bouquet of flowers of a certain type and colour, pinned to my jacket, to allow me to be accepted by his security team as a member of the official delegation.

During this airshow, I received one of the most disappointing messages in my life. While we were very active trying to sign a Letter of Intent with Malaysian Airline System and while the signals to materialise such document in the presence of Dr Mahatir were green in their vast majority, I received a phone call at 2 am in the morning from John Leahy, telling me not to sign anything, and to pack and come back home.

No one could imagine the disarray we were put in by this phone call. The reason behind this harsh decision was that *Airbus and Europe cannot be subject of retaliation from the US, for buying back the Boeing fleet in order to sell our aircraft*. But this was exactly what Boeing had done to us in two previous cases (in Singapore and in another country), just few months earlier.

This decision was taken by the Board and in particular by its chairman, the Aérospatiale CEO of the time. We were supposed to meet the airline the following day, to continue the discussions, but we politely suggested to give them more time and to meet again in the coming days or weeks. Needless to mention how they interpreted this change of attitude from our side, from speed up to slow down.

The funny side of the situation was, that while I was hiding because I said that I left the airshow, the chairman of the board showed up in Langkawi 24 hours later, came to the Airbus stand, and asked if there was something to be signed. You can imagine the laugh of the team present at the stand who was fully aware of the crazy situation.

This awkward event was at the origin of a song composed by the team (melody and lyrics), criticising the management. This song lasted for many years and was a sort of a rallying hymn for the team.

The last time I met Dr Mahatir was in October 2003, when I was invited to the inaugural, in Malacca, of the CTRM factory in charge of manufacturing some A380 parts, as per the offset agreement signed with Malaysia. He was attending the inaugural and had some nice words about me in his speech, when he mentioned his promise to me to buy the A380 provided Airbus accepts to manufacture some key parts of the aircraft in Malaysia.

He declared solemnly that since the airline needs a large aircraft and since Airbus has fulfilled its commitment, he will respect his promise to his friend

Habib, and MAS will buy the A380. This was a strange moment, because I was happy with this declaration and proud of the sale of the A380 to MAS, but I was not any more in charge of sales within Airbus, as will be explained later.

We had a very warm discussion after the ceremony, and we exchanged some words in Arabic, commenting on how will and efforts remove all obstacles. This moment was immortalised by a photo.

I was lucky to have met the two charismatic leaders of Malaysia and Singapore who were, initially, in 1963, citizens of the United Federation of Malaysia and Singapore, before the two countries split in 1965. Both were visionaries dedicated to their people.

They transformed their countries, putting them in the front wagon of the Asian *Tigers* train. More specifically, both of them bet on aviation to boost the economy of their respective countries, and both of them succeeded in this challenge.

During my different trips to Asia, I had a good opportunity to compare the point of views the British, French, Germans, and Spanish had of the different countries of the region. I noticed that there were quite some differences, due mainly to the culture and to the colonial experience of each country. My personal point of view was always slightly different from the European one.

My new role as Sales SVP was a difficult one; I was exposed to conflicting pressures from different sources. There was pressure from my CEO who had (sometimes) his own strategy, influenced by his experience in Matra and the advice of his close circle, as well as from my direct boss, the COO commercial who liked to be the number one salesman.

There was also pressure from the heads of regions, who liked to keep a direct connection with the COO commercial and run freely their region without any supervision and finally there was also pressure from some partners for whom Airbus was the place to be in.

The first blow to my leadership came by end 2000 when, following a lunch with all the heads of regions and attended by the CEO and the COO, each head of the region had to give his appreciation of how things were working in the sales division. Apart from two persons who were requesting more freedom for operations (because I had refused some of their proposed strategies), all the rest of the group focused mainly on logistics issues and on financial aspects, including their salaries.

I never understood the reason for this lunch, but its outcome was very surprising, since all the head of regions were promoted to the position of Senior Vice Presidents (the same level given to Christian Scherer and I).

As of that date, the COO took full control of the sales teams and de facto sidelined me. I kept control of Africa, Middle East, and Europe, while Asia and South America fell quickly under the direct control of the COO. The other regions remained a grey area for some months.

This situation was not helpful in the middle of our efforts to beat Boeing and become number one (indeed, we started matching Boeing).

Now, I must describe briefly the interaction between Airbus and the partners since the arrival of Matra. Before, each partner could be called upon for help, either at the sales level or at the financing level (to get the guarantees of the credit agency of its country). Aérospatiale, BAE, CASA, or DASA were not interfering in our daily business, but were called upon if and when needed. But from the four, two were generally more active than the others.

These were Aérospatiale and BAE, thanks to their subsidiaries spread all over the world, and also the leading role Coface and ECGD (now UKef), their credit agencies, were playing in securing financing for some difficult customers. But since the arrival of Matra, the new partner Aérospatiale-Matra seemed to want to play a leading role in the G.I.E.

Sales were the visible part of the commercial organisation and of the company in its entirety, and lots of people liked to be involved. The slogan of the commercial symposium *nothing happens until somebody sells something* made a lot of people think that, in order to exist and get promoted, you must show up during sales activities.

In particular, the situation became more complicated with the merger of Aerospatiale with Matra, just a few months after the arrival of the new CEO, Noël Forgeard. It was announced in November 1998 and became effective in early 1999.

As a result, Aérospatiale (so far owned by the French government) merged with Matra which was owned by Jean Luc Lagardère, who became the chairman of the company. The new entity was to be run like a private company listed on the stock exchange.

In this respect, I never knew if Noël Forgeard was aware of the upcoming merger. Some sources say yes, he was aware, and he was sent as a *submarine* to pave the way for the merger. Others say he was not aware and was surprised, at least initially, by the move. They insist that he left Matra following the failure of the merger project with Thomson-CSF.

His friend, President Chirac, then is said to have intervened with the CEO of Aerospatiale and had suggested that Noël Forgeard be appointed head of Airbus, in replacement of Jean Pierson.

Nobody has the real answer to this besides him, the late Jean Luc Lagardère, and a very limited number of persons close to the merger negotiations.

It must be said that Aérospatiale and Matra, as companies, had nothing in common. But as soon as Lagardère took a large stake in Aérospatiale, and after the two companies merged, Aerospatiale-Matra became one of the four partners of Airbus Industrie G.I.E., and through their international commercial arm, AMLI (later SMO, or Strategy and Marketing Organisation), started trying to have a say in the Airbus' commercial strategy.

Since I was promoted, I could feel the multiple attempts of AMLI/SMO to get involved in the sales campaigns. A year, after the arrival of the *Matra boys*, as we used to call them in Airbus, I was invited to the yearly conference of AMLI and was surprised to discover that each region of AMLI was reporting its commercial activities, including those related to Airbus, as if they were undertaken by AMLI.

I did not understand the new setup and continued to work without any special coordination with the partners, besides the usual requests for some political lobbying through the respective governments.

There were some possible areas for close coordination, like harmonising the offset obligations of the different divisions within the partners with those of Airbus to maximise the effect and minimise the cost and ensure full compliance with the new EOCD rules, applicable as from October 2000. For these important subjects, we, at Airbus, worked alone.

For example, Airbus was the first to become fully compliant with the new EOCD rules, and we made a roadshow to present our new policy to our partners but did not get any reaction from them. For offset projects, it took years before a serious coordination was put in place.

Here, again, I got the confirmation that visibility during the sales campaigns was important for the management of some of our partners (you must be seen as part of the success and avoid being seen as part of the failure).

The aircraft deals and mainly the big ones are glamorous and attract the attention of the media and the politicians. If you are in the photo of the signing ceremony, you will make headlines for a while.

Thanks to our sales successes, we could work relatively independently, but one could feel the pressure. I had no problem to coordinate more with AMLI, but I felt there was an incompatibility of strategies.

The Airbus guys were used to work with civil products, with civil customers which could be privately or government-owned, and with several of them in the same country.

The Airbus people were working in small teams with differing types of expertise for long periods of time, and using multiple and repetitive studies covering performances, technical specifications, and economical aspects. They were dealing with a limited number of products which they knew perfectly. Political lobbying had a very limited impact, and in most cases, there is a long-lasting relationship with the customers.

On the other hand, the Matra boys were used to multiple military products, with one customer per country: the Ministry of Defence. Hence the need for strong political lobbying. Their sales campaigns were, usually, much shorter in time, and they needed less studies to convince the customer. It is not a criticism; it is, simply, a fact imposed by the reality of the business.

This difference in the commercial approach explains the divergence in the commercial strategies. In addition, the Matra boys found themselves dealing with additional products from Aerospatiale like helicopters, satellites, and a lot of military hardware.

They were genuinely convinced that their role was to coordinate the sales activities of all the divisions, but while the Airbus teams dedicated 100% of their time to sell their products, the AMLI teams could only dedicate a portion of their time to sell Airbus products. They could assist in certain cases, but they could not substitute.

I can say that 1998 was a good year. On top of my new job, I met many interesting people besides the politicians, and my team scored many sales successes. Among the new people I met, I must mention, with strong focus, Mohamed El Hout, the newly appointed chairman of MEA. Following the end of the civil war, MEA went through some difficult years of rebuilding and restructuring.

Late charismatic chairman Selim Salam passed the helm to a successor who had difficulties to get the full political backing badly needed to manage such a complex organisation as MEA. It took a few years until the central bank (Banque of Lebanon), which de facto owns the company, decided to take full control of MEA and to appoint one of its young executives as chairman, Mr El Hout.

I met him for the first time, in Tunis where he was attending an AACO (Arab Air Carriers Organisation) general assembly. We found many common points of interest, and thanks to very friendly and nice comments made by other presidents and chairmen about me, he came to see me and asked me if I could help him rebuild his airline. From that date, we became best friends, and still are.

Thanks to this strong mutual trust and to the ability of Mohamed to master all the airline business elements which he learnt in a very short period of time, to his frank and straightforward approach in dealing with people and companies, and to his visionary business conduct, Mohamed and his team, and particularly his right hand, Yassine Sabbagh, succeeded in rebuilding MEA and to overcome all the heavy storms the country and the airline had suffered during the past 25 years.

For all his achievements, he largely deserves to be recognised as one of the best airline leaders in the world.

It is already difficult and challenging to manage an airline in any country, but in a country which suffered several wars causing the bombardment and the closure of the sole airport, where any decision requires the approval of several political factions, and where a serious financial crisis added to a catastrophic accident put the country on its knees, managing a national carrier can be considered as an impossible job. Mohamed El Hout did it and is still doing it.

Hats off to the artist.

On 10 July 2000, EADS (European Aeronautic Defence and Space Company) was born, through the merger of Daimler-Chrysler Aerospace-DASA) from Germany, Aerospatiale-Matra from France, and Construcciones Aeronauticas Sociedad Anónima (CASA) from Spain. Its capital is split between DASA, Lagardère, the French state, the Spanish state, and one-third floating.

This new change induced a complete reshuffle of the management with the arrival of Matra or Lagardère boys in Airbus and in the other divisions (Helicopters, Defence and Space), and at corporate level. We also saw the arrival of some Germans and a few Spaniards (balance requirement) in key managerial positions which had become really *political*.

I hoped for an improvement in the situation, especially as I was directly involved in the launch of the A380, and I got the first commitment signed. Late Mr Lagardère was following the A380 closely, and further to my first meeting with him, I got a comparatively positive impression.

I succeeded in managing the whole situation for two years (before and after the creation of EADS). Thanks to tremendous teamwork and despite all the internal disagreements, Airbus became number one in sales. Of course, it was not my prowess alone; it was the cumulative result of all the commercial team efforts.

In parallel with our excellent commercial results, I noticed a growing trend to limit my control of the sales teams and the sales strategy. It started with the acceptance by the CEO and the COO to give more freedom to the heads of regions and to promote them (as explained previously).

This was followed by two to three cases where the CEO and the COO ignored that I had basically closed some deals with some customers, and, despite that, engaged in additional negotiations, with the sole result of ending up giving further price reductions which were not needed. I never understood the reasons for such a behaviour.

If it was about the paternity of the deals, I never tried to brag about the sales achievements and always left the signature of the contracts and the announcement of the deals to the CEO or the COO.

There was an instance when the CEO travelled to Libya, with a full delegation, to meet the leader Kadhafi, without informing me. I had, previously, explained the situation in Libya and warned that some people, from or close to the government, were trying to mislead the Western countries and lure them about possible business in the country, thanks to direct contact with Kadhafi.

One of them succeeded to lure some Airbus people. Knowing my position and my warning, the easiest solution was to sideline me.

The trip took place with one of the test aircraft (an A340), and it ended in complete failure.

For a few days, none of the members of the delegation dared speak to me about the bizarre trip. There was a part of shame and a part of fear that I escalate the matter. Libya was very sensitive, and any blunder could be catastrophic.

By mid-2001, the situation became very difficult to manage, and I dared to criticise what was happening. This led to a very strange discussion I had with one person from the inner circle of Noel Forgeard. The person invited me for a drink in Paris, at the Warwick Hotel, near les Champs Élysées, and started speaking about different subjects, outside the business area.

It happened that we had common interests in some scientific matters like quantum mechanics and the string theory, and we used to enjoy exchanging opinions on these subjects. I was puzzled, but at the same time pleased that this gentleman finds time to invite me for a coffee and debate, with me, about interesting scientific matters.

Unfortunately, this impression was completely destroyed at the end of our meeting, when he asked me, with a friendly tone, to avoid contradicting or criticising the CEO, in the future.

I immediately understood, from this message, carried to me in a Mafiosi way, that the game was over. Consequently, I decided to resign, and I wrote a letter to the CEO and the COO, in June 2001.

The answer of the CEO was verbal:

"Why this resignation? How am I going to explain this to Mr Lagardère? We are overtaking Boeing, we are in the middle of the launch of the A380, and my head of sales is resigning."

I simply answered that I was not anymore the head of sales and that I'd still fulfil my commitment to achieve the launch of the A380. This exchange was followed by a meeting with the CEO, the COO, and the head of HR.

During this meeting, a solution was found based on a study I had done, about the development of *services* which would become an important source of revenues. It consisted in creating a subsidiary, abroad, to sell Services, preferably as a joint venture with an airline, and I should head this new subsidiary.

This solution satisfied all parties: the COO got direct control of the sales teams (no head of sales was appointed after my departure), the CEO would not have anymore somebody to criticise his strategy, the head of regions would have more freedom to operate; and as far as I was concerned, I had nothing more to prove in terms of sales successes. It was time for me to try other challenges far from headquarters.

For different good reasons, Emirates was the best choice to partner with in the joint venture, and Dubai, the best location (geographically and logistically) for the service company, which we called *Total Airline Service Company* or TASC.

As of July 2001, a memo was issued to appoint me in the new position, and I spent the remaining part of the year managing the sales teams and preparing the setup of the new company in Dubai, while negotiating personally the first A380 contract for 22 aircraft.

After the signature of the A380 contract with Emirates, and the Letter of Intent for the setup of a joint venture in charge of selling services to the airline community, I settled in Dubai with the plan to stay there for five years. For several consecutive reasons, I am still living there. I can say that with Tunis and Toulouse, Dubai is the third mark in my timeline. I had a first part of my family life and career in Tunis, a second one in Toulouse and a third one in Dubai.

This being said, I have to stress the fact that the interactions with the UAE started long before (1988) and that some of my major professional achievements and even some of the most important family events (wedding of my son, birth of my grandson) took place in the UAE. Hence a very special connection with this country.

TASC (Don't You Forget About Me-Simple Minds)

As explained later, TASC was finally set up as a simple 100% subsidiary of Airbus, during the first half of 2002. Consequently, I recruited the core team, and we started our operations by July 2002. TASC was a very interesting adventure because it brought me back to the airlines' environment. I had the good side of both worlds: the airlines and the manufacturers.

In order to sell services to airlines, you have to think and react as an airline manager, and you have to cover all the aspects of airline operations, from commercial to handling, to flight operations to passenger services, to maintenance and financial services.

For this reason, we hired people with airline experience, and we developed dedicated services, ranging from setting up a new airline (business case, obtention of Air Operator Certificate—AOC, network and fleet planning, etc.) to yield management, to cabin interior design, to used spares trading, and management of the supply chain, etc.

TASC was relatively successful and was able to bring some innovations such as the design of a new business class concept which became the standard in the industry.

The main reason behind the creation of TASC was the fact that the Services activity was growing, and was becoming more diversified and more lucrative. All the studies were converging in showing that Services revenues were becoming close to those generated by the sale of aircraft.

While the engine and components manufacturers understood this development and were already boosting the sales of their Services through different concepts, like Power by the Hour (payment of a fee for each flying hour, against the maintenance of the engine), the aircraft manufacturers were comparatively late in following that trend.

Therefore, TASC acted as an accelerator for Airbus which discovered that Services represent a serious opportunity to enhance revenues.

Unfortunately, once again, this new activity developed by an Airbus subsidiary attracted the attention of the rest of the company and, at the same time, raised a lot of jealousy and negative reactions. The same people who were not able to improve the services to customers through the existing structures within the company, started pushing for the extension of the scope of Services handled by the Customer Support Directorate.

As a result, TASC started having a serious competition from its mother company. This was the beginning of the end, and it showed that success does not necessarily generate admiration, but rather envy. Another lesson learnt for me regarding human relations and behaviours. So sad.

During my days at TASC, I had two serious health problems. One in March 2004, which was close to kill me: a deep vein thrombosis, causing a pulmonary embolism and a clot in the heart, and another one in May 2005, which was an epistaxis (or serious nose bleeding) resulting from the medication I was taking.

I'll never forget the support and care I had from the Airbus family. I always remember Noël Forgeard standing at my bedside at the Clinique Pasteur in Toulouse, at 7 am, the day after my arrival there and asking Professor Jean Marco:

"Can you please save him?"

This was Noël. He could also be a good friend full of humanity and compassion. My relationship with Noël was very friendly and full of mutual respect. But as I had dared tell him the truth as a friend, some people from his entourage misled him and led him to interpret my behaviour as a lack of respect, which was absolutely not true.

I fully respected Noël and had very friendly feelings towards him and his family, and I believe the reciprocity was also true. My only problem was there were many hurdles in front of Airbus, and I felt I had to give him timely alerts.

Unfortunately, the communication and image aspects were leading the show, and his entourage was adamant to discount all bad news. Noël must shine in a cloudless sky. Everything must be done to promote him to get the position of Head of EADS. This was the objective, and nothing must get in the way and stop it.

Like a few others—but they were very few, I was a sort of Cassandra, and did not fit into the scenario. I was needed because of the good sales results, but I was not fully playing the role of a Noël promotor. I feel no grudge and no frustration. This is the real life in the majority of the corporations. Again, a new lesson learnt!

I had with Noel another memorable and friendly interaction in June 2004, when he gave me the decoration of knight of the French Legion of Honour (Chevalier de La Légion d'Honneur).

This decoration had been granted to me by former President of France, Jacques Chirac, following a proposal made by a group of aviation professionals led by my friend Fernand Danon who is a Tunisian Jewish pilot (he was working in France and is also holding a dual French and Tunisian citizenship), endorsed by Gilles De Robien, the former French Minister of Transport, and supported by my late colleague Nicolas Girod who was in charge of government relations at Airbus and whom I had hired years earlier as a salesman.

Fernand Danon was a strong supporter of the A380 and believed that I deserved a certain recognition from the French authorities, for all my efforts to launch and to sell the A380 (by late 2003, with my team we were at the origin of the sale of close to 45 A380s out of a total of 112 aircraft sold. Finally, we achieved a net order of 143 out of 251 up to 2016).

He did everything discretely and informed me once his letter was in the hands of Minister De Robien. I highly appreciated his gesture, and I was happy to have a Jewish friend supporting me. This was a brilliant expression of the spirit of tolerance which prevails in Tunisia.

The Légion d'Honneur had an additional special taste because it was granted to me as a French citizen, while I had not satisfied the minimum requirement of 20 years of French citizenship. Gilles De Robien had requested a special derogation from President Jacques Chirac to grant me the Légion d'Honneur as a French citizen.

This was swiftly approved, and Nicolas Girod informed me of the comments the different people involved in the process had made: they were positive, supportive and full of gratitude.

This made me forget the previous negative interactions I had had with some French people. In addition, I was receiving my medal seven decades after my father received his glorious *Médaille Militaire* after the liberation of France at the end of WW2. Somehow, I was closing a loop in the relation my family had with France.

Noel delivered an emotional speech, full of heart feelings, while not forgetting to mention my recent health problems.

This was the last happy moment I spent with Noel.

My Légion d'Honneur also generated some jealousies and a few discrete requests were made, during the following months, from some executives to Noel, to get the same decoration. As from late 2005, we could witness a number of ceremonies related to either *La Légion d'Honneur* or *l'Ordre du Mérite*.

I left TASC in June 2006 to come back to Airbus. My deputy tried to continue the adventure, as long as possible, but the new situation within Airbus put strong limits on TASC's ambitions and the company was closed in 2012.

From this enriching experience, I learnt that creativity and innovation are not always welcome, and it is not to be excluded that these could face serious opposition or even sabotage from the closest people. I also learnt that there are always a lot of sharks in the market, pretending to sell their services.

But either they don't have the necessary resources to be able to deliver what they promise and deliver very low-quality work, or they use their company to lure customers and act as a simple go-between. The fact is however that being a subsidiary of Airbus with its strong brand, helped a lot to give TASC a good image in the market.

I was happy with the TASC achievements, materialised by the fact that many top-quality airlines entrusted it with some critical tasks, like restructuring their commercial activities (such as, for example, MEA) or designing the cabins of their new fleet of Airbus and Boeing aircraft (Etihad).

One other consequence was the strong push for Services, operated within Airbus, which led to a completely new organisation following the demise of TASC, which served Airbus, dead and alive.

Concerning the relations with Etihad, it is worth mentioning that I was among the first witnesses of the birth of the new company. Sheikh Ahmed Bin Seif Al Nahyan, who is a close friend and whom I knew for a long time since he was the CEO of Gulf Air, told me about the project and asked for my opinion on some key elements, like the name, the logo and the commercial strategy.

I started hearing about the project in 2002 when he told me about the intention of the government of Abu Dhabi to launch a new airline. The reason was that Emirates and Gulf Air combined (Abu Dhabi was still a member of the Gulf Air consortium) did not respond to the needs of Abu Dhabi in terms of volume of air transport (passengers and cargo).

Consequently, I followed the evolution of the project on which I had the opportunity to give my opinion, informally, to Sheikh Mohamed Bin Zayed Al Nahyan in the presence of Sheikh Ahmed Bin Seif Al Nahyan, during a wedding.

A brief discussion took place, and Sheikh Mohamed said that he was convinced and added:

"In sha Allah, we will proceed."

The airline was launched in July 2003. Sheikh Ahmed Bin Seif was appointed President, a few aircraft were leased, and the airline started operating in November 2003. A contract for the purchase of A320s, A330s and A340s aircraft was signed with Airbus in 2004; another contract was signed with Boeing for the purchase of B777, and TASC was appointed as the consultant in charge of designing the cabins of the new aircraft (Airbus and Boeing).

This was a *premiere*: an Airbus subsidiary working on Boeing aircraft and exchanging documentation with the competitor of the mother company. The new cabin was so innovative that it attracted the congratulations of Sheikh Mohamed Bin Zayed and Boeing.

Despite living in Dubai, I had to shuttle several times per month, to meet Sheikh Ahmed Bin Seif and ensure that the new cabin was working in a satisfactory manner. From then on, I had two main airlines in the country to look after, each one headed by a Sheikh Ahmed. Luckily, the two were and still are good friends and I would have many opportunities to meet them together.

The come back to Airbus.

As explained later, I was called back to Airbus in 2006, to oversee a new subsidiary called Airbus Middle East, based in Dubai. Consequently, I kept a supervising role in TASC, as chairman of the board. This return took place during a period of turmoil within Airbus and obliged me to manage several crises, some of them linked to our big customer Emirates and others including Etihad and Qatar Airways which was also becoming a big customer.

The new Middle East subsidiary excluded Tunisia (as in the previous Middle East region) and also Libya (following the failed trip some years ago and my resignation in 2001. The management was happy to put Libya in the African region under the leadership of a newcomer).

The first crisis was with Etihad, which appointed a new President just around the same time I took my new function as President of Airbus Middle East (AME), to replace Sheikh Ahmed Bin Seif. For different reasons, the new CEO had a difficult relation with the company which was in charge of the maintenance of the Etihad fleet, and which was linked to the authorities of Abu Dhabi.

He did not renew the agreement with them. At the same time, there were some reliability issues with a part of the fleet. These problems were easily detectable and reparable if the aircraft were inspected as per the set schedule. Unfortunately, there were a multiplication of technical problems, most of them beyond the control of Airbus.

To show that he was a sharp manager and a strong leader, the new CEO asked Airbus to compensate Etihad for the drop in dispatch reliability by paying some amount of money, arguing that the technical problems were due to Airbus' negligence and had caused a loss of revenue to Etihad.

I did not agree with him, and we reached a deadlock. Unfortunately, the Airbus management decided to send one of the top executives to discuss the matter with the new CEO, and he accepted to pay the requested amount of money, contradicting my opposition and the reality of the situation.

Consequently, the new CEO called John Leahy and told him that he did not want to deal with the Airbus Dubai office anymore. I was expecting this move, because the new President was supported by a group of persons in Abu Dhabi who did not appreciate the fact that I was close to Sheikh Ahmed Bin Seif, who was appointed chairman of the board of Etihad, and that I had good relations with the leadership of Abu Dhabi.

Such situations are not that uncommon – everybody tries to negotiate as best one can – but usually responsibilities tend to be recognised for what they are, without further harming the relationships.

Despite my explanations, John Leahy accepted the request and appointed somebody in Toulouse to take care of Etihad. For me, this was, de facto, the beginning of a trend to diminish the credibility and the role of AME. Here again, John mentioned the notion of emotions, because I had said loud and clear that I could not accept a brutal and unjustified change of position vis-à-vis a customer, without prior consultation and without an agreed approach to explain the change.

I insisted on the importance of protecting the credibility of the Middle East team. I started being fed up with these unjustified comments which were real nonsense. Years later the top management acknowledged that they did a mistake when accepting the request of the new CEO, but, by then, it was too late and the caused damages could not be repaired, from all points of view.

I have already mentioned the problem with Qatar Airways which led to a similar conclusion concerning my involvement. In both cases, time showed who was right and who was not.

Concerning Emirates and due to the ongoing interactions with this airline since 1988, and the effects of these interactions on Airbus and on me personally, I preferred to put all the related events in a dedicated chapter.

I'll come back to some follow-up events related to Abu Dhabi, later.

5.3 Airbus Group (Major Tom) (Space Oddity-David Bowie)

After Tom Enders became president of Airbus in 2007, the situation changed tremendously. First, Tom was very quick in assessing the role and contributions of the different executives around him. He set clear priorities and empowered each one, according to his or her competence and effective contribution to the company. As he often said, he did not care about *the Bullshit*.

Second, he decided to make a tour, with me, of the main customers, in the Middle East region, one month after his arrival. We visited Egypt, Lebanon, Qatar, and the UAE within a few weeks. Third, he got clear messages from the visited customers on how they saw their interactions with Airbus.

After this tour, he gave me his full confidence and trust, and our relations became very friendly.

After Tom's arrival, John Leahy, in the frame of a complete restructuring of the commercial organisation, following the departure of some key executives, decided to give me back the responsibility for the corporate jet activities, and the return of Libya to the Middle East region, in May 2008.

Since my relationship with Louis Gallois, the president of EADS, was also good, I could rely on two solid supporters at the top of the company. I knew that, in case of arbitration, they would be fair and transparent. Louis Gallois used to call me regularly for one-to-one working lunches, during which we could exchange on all matters without any inhibition.

Despite some occasional disagreements with some colleagues, I enjoyed a very encouraging and supporting working environment. I also appreciated the fact that, on several occasions, Louis and Tom took good note of what I was proposing.

I must say that I enjoyed working with this fantastic duo very much. They were different but very complementary to each other. I never had any comment from Tom about my regular reviews with Louis, and I never heard Louis contradicting decisions made by Tom.

After the Berlin Airshow, in 2010, where we signed the contract for 32 A380s with Emirates, Tom called me into his office, and we exchanged some ideas about possible improvements regarding the Airbus organisation. He was not happy with some functions and had some ideas he wanted to cross-check with me. I took this discussion as a strong sign of confidence and did my best, first to do my job and second to help Tom with constructive advice.

One of the first possible improvements discussed was how to boost the sales of the A380.

Different scenarios were envisaged, but my comment was:

"How is it that the rationale set by Emirates is not applicable to airlines like Lufthansa, Air France, British Airways, Cathay, Singapore, or the main Chinese carriers?"

There was a paradox: all these airlines were trying to develop the mega hub concept and at the same time, they complained about the slot limitations at their respective hubs (Heathrow, CDG, etc.). Of course, there was the Boeing push to develop the secondary routes (which Airbus had actually been first in promoting when launching the A340 for long thin routes), but this should not be enough to block the A380.

I suggested, as I did with John Leahy, to review completely our marketing argumentation for the A380, using the elements tested and successfully agreed with Emirates. In addition, I proposed to put an updated version of the A350 engines on the A380 to achieve more than 10% in fuel savings.

Following a dedicated review made at the level of John Leahy, a dedicated team was eventually set up to market the A380, but they continued with the same old arguments. The engine upgrade option was rejected internally despite a strong insistence from some airlines like Emirates.

It is true that the market had changed tremendously and that despite the congestion problems caused by the big hubs, the airlines were shy to use the

A380 to solve these problems. This paradox deserves a more detailed explanation and still constitutes one of the main issues to be dealt with in the future.

At the end, the management had to make a decision based on several considerations, but essentially financial ones (without neglecting the fact that some executives were not fond of the A380 and preferred the more easily saleable aircraft like the A350 or A320). The sad result was the end of the A380 programme.

Another project I already mentioned was discussed and approved by Tom. It was the setup of a dedicated subsidiary for the corporate jet.

I was also called upon by Fabrice Brégier, who succeeded Tom as the CEO of Airbus Commercial Aircraft in 2012, to help him find solutions for several serious problems Airbus had with Mubadala/Etihad (Abu Dhabi) related to Strata, the aircraft components manufacturer, which was set up in the frame of Airbus offset obligations towards the Abu Dhabi authorities. These problems became very serious in 2013 and caused a serious disagreement between Airbus and Strata.

Even though Airbus Middle East was not in charge of the sales activities to Etihad, I embarked on lengthy and difficult negotiations with Strata, Mubadala, and Etihad. These negotiations lasted for months and allowed us to find acceptable and fair compromises. But I discovered that, for a while, Airbus was giving up a lot of things to Strata/Mubadala/Etihad for free.

It seemed that the management of Etihad had several times threatened Airbus of cancelling totally or partially some orders (particularly A350 orders). In reality, the intention of cancelling had been there for some time and was finally executed. For some reason I don't know, Etihad was looking for an excuse to cancel some orders.

This was a Damocles sword hanging over the head of Airbus for a while. When I heard about this continuous threat, I told the Airbus management that we must accept cancellations and accept that some airlines will never buy an Airbus product. This would not change our position as world leader. But if we act under the pressure of continuous fear of facing a cancellation, we could make bad judgements by trying to solve short-term issues without guaranteeing the long-term (a threat of cancellation can disappear for a while and reappear later).

I had to put things on track and appreciated the fairness of the Strata/Mubadala executives with whom I negotiated and who accepted some amendments, making the agreement more balanced.

I understand that Airbus has multiple relations with Abu Dhabi (military, space, civil aircraft) which deserve a much more global approach. I always appreciated the confidence placed by the authorities of Abu Dhabi in Airbus products, and occasional compromises are certainly necessary to maintain the quality of the relationship. But a continuous threat of cancellation coming from the same person over a number of years cannot be accepted. I was happy to notice that I had convergent views with the management of Mubadala and Strata.

One of the things which surprised me during these difficult negotiations was the fact that the Airbus team in charge of offset sent me an e-mail suggesting to create a training joint venture with Etihad. We were still stuck in a conflict about an already agreed project and had not progressed on a training joint venture requested by Sharjah based Air Arabiya for some time, although this was a better choice for different reasons (logistics due to the proximity of Sharjah to Dubai, international coverage thanks to Air Arabiya subsidiaries abroad, and existence of very competent instructors with whom we have already cooperated).

I told Airbus about the promise to Air Arabiya and warned them that Etihad was not the right choice and that it would not support the project because they have other ideas and priorities.

That was exactly what happened, and Etihad refused the project. This involvement with Etihad was an eye-opener to Fabrice Brégier, who understood that Airbus made a mistake in accepting the request of the CEO of Etihad not to deal with the Middle East team.

He understood that a close knowledge of the airline, the people, the country, and the culture was vital to understanding the reality and making the right decisions. A European person, based outside the UAE, could not have a clear picture. He acknowledged this in front of me. He had had nothing to do with it, but it was a real gentleman's behaviour.

On another aspect, Fabrice let me deal directly with Tom, who by then was the EADS CEO, on several subjects without any interference.

At the group level, we can say that the most important revolution brought in by Tom was the total restructuring of the company, when he became the sole CEO of EADS (up to that point there were always two, one from Germany and one from France), as of 31 May 2012.

During the summer of 2012, Tom and I had a long dinner in Hamburg, following the delivery of an aircraft to a customer. We were staying at the hotel Louis C. Jacob in Finkenwerder, and the plan was to go together to Munich the next day, where the head of the defence and space division needed my help to better understand the then fluctuating situation in Egypt.

As the new boss of EADS, Tom had some important meetings in Munich, and since he was the one who had initiated this meeting for me, he proposed that I travel with him on the same Lufthansa flight. Over dinner we discussed in detail the organisation of the whole group, and he suggested to streamline everything and put in place a new organisation which would take care of the simultaneous processes of globalisation and regionalisation of EADS.

We also debated about the role of the Strategy and Marketing Organisation of the group (SMO). He asked how the competition and groups like BAE and Rolls Royce were organised. I was lucky, because not long before, I had attended a conference about the future of aviation and how the manufacturers must adapt to the new technologies and challenges for the future.

On that occasion, some manufacturers had presented their organisations and how they had made some necessary adaptations to prepare for the future such as, for example, to replace some functions with artificial intelligence. Thanks to these presentations, I was able to suggest some ideas to Tom.

I know he had also been advised by some other executives to bring in some changes. Some internal work had been processed by his cabinet about different reorganisation options for several months.

In the summer of 2012, Tom was also busy with a project of merger between BAE and EADS, to create a strong pole of European aerospace which would have seen the centre of gravity of the European aerospace industry shifting slightly closer to the Channel.

This project was strongly opposed by Germany and the US, and was finally dropped. Following this failed attempt, Tom engaged in a serious restructuring of the group.

By mid-2013, he invited me, with three or four other executives, to a working lunch at *Le Cercle d'Oc*, a restaurant close to the Airbus headquarters in Blagnac, to discuss how to redefine the role of SMO, and what the priorities for reorganisation were. One of the executives attending the meeting was tasked to draft a new job description and organisation for SMO, in the frame of an integrated group.

During the same summer, we spent a few days together with Tom, in my house in Hammamet in Tunisia. I avoided speaking business with him, but he provoked me, on the third or fourth day, by asking me what my suggestions would be if I was in the shoes of the gentleman in charge of making the proposal.

I told him bluntly:

"You are fond of the US. Then just replicate what they do but let us revive our pioneering spirit. First, remove this stupid name of EADS and make it simply Airbus, and second, make it a single umbrella for a really unified company."

He listened to me carefully, smiled and nodded, adding:

"This idea of changing the name has been in my mind for a while, I already discussed it with some executives and it could be the key to trigger all the changes."

By the third quarter of 2013, the report was submitted to Tom. He did not like it because it did not respond to all his expectations.

Nevertheless, he called me, one day, in his office and told me:

"You are going to play a role in this new organisation. You are going to implement the concepts of one roof, one voice, in the frame of the globalisation and regionalisation of the company, in a big region regrouping Africa and the extended Middle East. You have to prove that it will work."

A few months later, in 2014, Tom announced the change of the name of EADS to Airbus Group and a complete reorganisation of the company. He also announced the regionalisation with the creation of subsidiaries in each region, acting as a single roof for the whole group (the divisions as well as the central functions) and speaking as a single voice on behalf of the whole group (the concept of one roof, one voice) and informed me officially that I was appointed president of Airbus Group Africa & Middle East (AGAME), and that Marwan Lahoud, the head of SMO, would soon formalise the corresponding decision.

Marwan Lahoud is a brilliant engineer who had joined Aérospatiale in 1998. He was instrumental in many key decisions linked to the merger of Aerospatiale with Matra and to the further restructuring of the new company. We had friendly relations, and our origins contributed to creating easy communication channels between both of us.

I really enjoyed our discussions about mathematics, quantum mechanics and, in particular, about the string theory. I always admired his sharp intelligence and his great culture. One of our favourite subjects of discussion was the Phoenician civilisation and the history of Carthage.

Despite our strong proximity, it took up to October 2014 to see my appointment materialise. Nevertheless, I did not waste time and prepared the structure by pre-selecting the different key managers to help me run the show. Therefore, I was able to start running as soon as 1 November 2014. Within six months, AGAME succeeded to show all the benefits of the new organisation, and Tom did not hesitate to use it as a model for the other regions.

One of the major achievements was the ability of AGAME to create, with investors in the region, new channels for financing Airbus aircraft sales, particularly in Saudi Arabia. The creation, with private investors, of ALIF, the Islamic financing fund, was a premiere and was praised by many specialists. Yann Ballet, the CFO of AGAME, was instrumental in the creation of this fund with its two co-managers, Idriss Ghodhbane and Moulay Omar Alaoui.

As much as ALIF was a success, through the financing, via a leasing structure, of 55 aircraft to be delivered to Saudia Airlines and Kuwait Airways, it nevertheless drew scepticism and criticism within the newly created Airbus Group, to the point of triggering a full audit from the company, conducted by an external US law firm.

The audit concluded that the fund and all the related transactions were sound. But this did not prevent the negative attitude towards AGAME to spill outside the company, through carefully planted articles in some media in which ALIF was described as suspicious, where AGAME was reported to be audited, and finally where my name was directly mentioned as the one who was helping Tom dismantling SMO (*Fekih Corporation* versus *Bullshit Castle*, as Tom used to call the building in Paris where some central functions of EADS were located).

I did not need to be a genius to find out that all the information reported by the media was provided by some people from inside the *Bullshit Castle*. I know who they are and why they did this. As I said at the beginning, my origins, my personality, and my background were and will never be accepted by some people, whatever I do. So be it. But it did not stop me to do what I thought was right to do, and it did not change me.

On the contrary, it strengthened my will and desire to defend my identity and my values.

When Airbus started the investigation into some potential corruption issues, I was also the subject of some suspicions concerning two or three files in which I had, de facto, never been involved.

For a few weeks, I was noticing hostile attitudes from some executives without understanding the reasons—until the day John told me that there were problems to finance some airlines due to the suspicion by the European Credit Agencies of the existence of possible corruption. As it is said in French, *je suis tombé des nues* (I fell from the clouds).

I had no clue about any of the mentioned files. Without hesitation, I went to Tom's office and asked for an urgent meeting with the Chief Legal, Ethics, and Compliance to clarify the situation.

Tom called the Chief Legal, and as soon as I laid out to him the situation, he declared to Tom that I had nothing to do with the mentioned problems. He added that, in one of the cases, the person responsible for the file had come to see him and explained all the details of what had happened. He insisted that I was all clear.

For the first time, I burst into anger and told Tom:

"Why this attitude of blaming me without even telling me and asking me for clarification?"

The Chief Legal answered:

"Unfortunately, some people told us it was you who did it."

This was the straw which broke the camel's back. Enough is enough! There was a very structured plan to tarnish my reputation and destroy my credibility for the simple reason that I had raised my voice in the past to criticise some errors, and now for helping Tom restructure the company and get rid of all the duplications and the unnecessary structures inherited from the past.

Luckily, all the audits and investigations showed that I had not played any role in any of the concerned investigated files. Even, when I had a say, in some others due to my responsibilities, it was demonstrated that there was no wrongdoing on my part.

One of the leading outside lawyers conducting the investigations told me after he concluded his report on one file:

"I am sorry to tell you that I have a strong impression that there are some people who like you to play the role of the scapegoat. I don't know the reasons, but for sure, they don't like you."

Another colleague, who has Arab origins, gave me a plausible explanation:

"Ils te font jouer le rôle de l'arabe de service." (They make you play the role of the *'serving Arab,'* meaning *'whipping boy'*.)

In this respect, he was not wrong.

During my more than 30 years with Airbus, I noticed that some people (luckily, very few) continue to be suspicious vis-à-vis an Arab or a North African employee. They were unable to accept that we could be as competent and dedicated as our European colleagues, working the same way as they do.

Thanks to Tom's trust, I succeeded in overcoming all the hurdles and helping him validate the concept of regionalisation. But as agreed with him from the beginning, I confirmed to him, by late 2015 that I'd retire at the age of 65, i.e., by the end of March 2017. Consequently, I planned all the next steps with the top management of the company, including choosing my successor, who was appointed as my deputy from the beginning of 2016.

The transition was very smooth, and I left as planned. Tom and Fabrice insisted on organising a farewell party at Airbus headquarters to celebrate my departure. They presented me with a very emotional gift—a forward engine mount from the first-built A380.

The part had an inscription saying:

"From Airbus to Habib Fekih" and the date of 10 April 2017.

I really appreciated the kind attention of Tom and Fabrice, who insisted that I had to leave in style—not to be confused with a number of other colleagues who had, unfortunately, been pushed out discretely. Some of them did really not deserve it.

I'll always remember the nice words of Fabrice and his warm and friendly speech:

"We must celebrate your departure. You deserve it, and I don't want people to think you were pushed out."

Nor will I ever forget Tom's statement to the media.

Despite all this, some French media did not refrain from describing my natural and planned retirement as a push-out.

Unfortunately, Fabrice himself eventually had to leave in a discrete manner, which I profoundly regret. He was among the few Lagardère boys who embraced the Airbus spirit and acted accordingly. He was competent, intelligent, and fair—a great professional.

I enjoyed working with him, and we agreed on many things. Despite his disagreements with Tom about some management issues and their competing ambitions, I never suffered from their difficult relationship. I was able to maintain friendly relations with both of them, and they were both gracious in respecting my relations with the other person.

In summary, I had three successive lives at Airbus:

1. The first one, for some 12 years, with Jean Pierson. It was the continuation of the pioneering adventure which started in the early 70s, with Bernard Lathière, Roger Béteille, and Felix Kracht. It was a period of brilliance, of continuous innovations (the A3XX/A380 was started during this period), of dedication of the employees, and of true friendship between all Airbus colleagues at the GIE, which counted no more than 2,000 employees by the mid-90s. Everybody knew everybody.

The working conditions were sometimes rugged and the salaries were modest compared to other industries, but everybody was happy and excited to work for the company in order to *beat Boeing*.

Jean Pierson was fully empowered by the four partners, who just retained a controlling role.

There were no prima donnas; egos were controlled, and no communication gimmicks were allowed. Communication was focused on the brand and on the products.

2. The second one, for about eight years, with Noel Forgeard: It was a completely new life, within a more structured corporation, with well-defined processes, a real board of directors, and the shareholders having the last say.

The working conditions were better, the salaries much better, and for the first two to three years, the enthusiasm built during the G.I.E. continued to prevail. The different improvements contribute, also, to boost motivation.

Airbus became the world leader and a regular feature of the world media, boosting the image of some key executives.

Unfortunately, the management style, the rise of some prima donnas, the extensive use of communication tools to promote the image of some executives, the tight control of the financial resources to satisfy the shareholders, and the increase in number of employees at headquarters to up to 5,000 led to a certain degradation of the overall atmosphere within the company.

People continued to work hard to achieve the company's and personal targets to guarantee their revenues, but there was no more fun. Customer satisfaction declined because the tight cost control caused quality issues with aircraft deliveries and less satisfactory customer support from the customers' point of view.

In addition, the company became shy of the usual trailblazer innovation track due to the shareholders' desire to minimise new investments, to compensate for the A380 substantially overshooting the initial development budget, and to ensure a minimum profit to get dividends (which is fair).

This second life was characterised by good commercial performances, good bonuses, and a dominant role in communications, with a tight control of the image of some executives and on how and when some events were to be disclosed and promoted. Everything was done to serve the purpose of some executives who had to shine.

The employees' motivation declined, as well as the fun aspect of the job. The situation became difficult with the A380 industrial problems and the consequential delays, culminating with a real crisis and with the inside trading investigation which rocked the company. But, thanks to a strong survival reflex which boosted the motivation among the whole personnel, these crises did not affect the daily running of the company nor its fundamentals.

We overcame them, and Airbus got out of them stronger.

Many shortfalls were corrected, but others remained.

Noël did not have the full power.

He was sharing it with the EADS Board and with SMO.

3. A third one, during 11 years, mainly with Louis Gallois and Tom Enders.

This was practically a continuation of the second life with, initially, some corrections of the main shortfalls and some improvements. But it ended with a revolutionary restructuring of the company, making the EADS group becoming a single global corporate entity called *Airbus*.

It is a streamlined company, with solid processes, efficient controls, and clear objectives (the Covid pandemic showed how the company could weather the storm and get out of it stronger than before).

Airbus became a global corporation, more professional, technically more efficient, and it is now the solid number one aircraft manufacturer in the world.

Tom had succeeded in creating a real world leader and given it all the necessary tools. Unfortunately, some of his top executives did not grasp the full meaning of this new setup. Some continued to work as before, and others did not understand that the prima donna era was gone.

Tom had more power than Noël, but less than Jean.

He could not finish his job as planned. The investigations, which started as early as 2015 and their multiple consequences, hindered his ability to fine-tune all the aspects of the new organisation, including the choice of the right people.

Nevertheless, Airbus started to ensure a more equitable balance between the three stakeholders: the shareholders, the customers, and the employees. Having retired from Airbus in 2017, I don't know what was achieved since.

It was a long evolution which had lasted more than 30 years, leading Airbus to become a strong corporation, but it lost its initial soul.

Is it bad? I don't think so. This is a normal evolution for any successfully growing corporation, and we have to adapt.

I feel lucky for having experienced all three phases in the life of Airbus. I had successes and experienced exciting moments in each one of them. It is true that the first one was the funniest and that it is much better to suffer while having fun than to suffer just for the money compensation. But people evolve, and maybe they prefer to have dedicated moments for the fun outside the working environment.

I feel proud for having given more than 30 years of my life to this company. I am proud of what I have done for Airbus and grateful for what it has done for me.

Above all, I feel honoured and lucky to have worked with the fantastic people of the different consecutive teams. I owe them my successive achievements, and I am grateful for their support and friendship.

5.4 The Englishman (Sir Tim Clark) (English Man in New York-Sting)

Since my student days, I had the opportunity to meet a lot of British people: English, Irish, Welsh, and Scottish. Some of them became good friends, and I share with them many things, but I consider only one of them as a very close friend because we have many things in common, and in particular, we have the same vision, the same approach, and the same ideas about aviation and its future. This gentleman is Sir Tim Clark.

Sir Tim was born in November 1949. He attended Kent College and is an economics graduate from the University of London.

His father was a ship captain, and in the late 50s, the young Tim used to travel at least twice a year, alone, from London to Southeast Asia and to other places for family reunions. He started getting the aviation virus from these multiple, long, and exotic flights.

His passion for aviation pushed him to join British Caledonian in 1972. Then he moved to Gulf Air, where he got acquainted with the Gulf region.

With several other Gulf Air managers and executives, he moved to Dubai in 1985 when Emirates was created, and he became one of the main pillars of the founding team.

I met Tim (I hope he will forgive me to call him simply Tim), for the first time in March 1980, in Bahrain, in the office of the late Mohamed Al Maskary, who was the head of planning at Gulf Air. Tim was a young manager in charge of market analysis.

I was just appointed as Director of Technical Development at Tunis Air, and I was tasked to look for possible technical cooperation with airlines in the Middle East. Consequently, the DG Technical and I embarked on a tour around several countries in the region. The first stop was Bahrain, and we received a warm welcome from the airline management.

Tim was briefly introduced to me by the late Mohammed Al Maskary, who was the head of corporate planning at Gulf Air, and we exchanged a few words. This was enough to give me a good feeling of the person. We were close in age, and we seemed enthusiastic to work in the field of aviation.

I lost sight of Tim for a few years, until my first visit to Emirates as Vice President of Sales Middle East in early 1989. Stuart Iddles decided to introduce me officially to Emirates before the planned visit to the airline with Jean Pierson. During this first visit, I met Sheikh Ahmed Al Maktoum, the chairman, the late Sir Maurice Flanagan, the Managing Director, Tim Clark, the head of planning, and several other managers.

It was a pleasure to meet Tim again, and it was funny to notice that we had done similar moves in terms of career evolution. Very quickly, we embarked on lengthy and interesting discussions. I came back a few days later, to prepare for the CEO's visit, and Tim was my counterpart. We had several discussions about many subjects: aviation, economy, music, sports, history, and politics.

We agreed on everything. I discovered, very quickly, that I did not need to listen to Tim's complete sentences. I could grasp what he meant from the first words, and I think the reverse was true. The visit with John Pierson allowed us to strengthen that bond and a clear mutual trust was born.

Emirates had already ordered a few A310-300s and A300-600s from Airbus, with some innovations in the cabin, like individual videos and Satcom telephone, and had introduced a new standard of on-board service. These consecutive orders and the deliveries of the first aircraft created friendly relations between the Airbus and Emirates teams.

A very unusual event strengthened these relations, let us call it the *dead rabbit*. During the delivery of an A300-600 in 1990, everything was ready for the final steps, when a very bad smell started spreading through the whole cabin. The joint technical team started a meticulous check in all corners of the aircraft, but without any success.

Finally, after a second round of checks focused on the air conditioning system, a dead rabbit was found in the aircraft duct which connects the aircraft to the auxiliary air conditioning packs on the ground. It was a relief to find the source of the problem, but the aircraft needed to be ventilated for hours before it could fly to Dubai. The problem was that it was Friday afternoon, and the French weekend was looming.

Finally, Maurice Flanagan, Tim Clark, and I decided to delay the ferry flight till Monday. Now, the problem was: what could we propose to the Emirates delegation for the weekend? Very quickly, a solution was found: organise a visit to some Bordeaux chateaux.

By Friday evening, we were in the Bordeaux region, in a beautiful old chateau bearing the name of one of the most famous wines. During these two days, the two teams interacted without any business constraints, getting to know each other, outside any protocol and without any inhibition.

This gathering, after a while, led to the creation of the *Toulouse Taste Vin Wine Club* with the initial participation of 12 people from Airbus and Emirates. Later on, more people joined, but the core team remained stable for many years. Each year, all the members pay a contribution of a few hundred US dollars.

Once the money is collected in due time, some members of the club go to the Bordeaux region to shop for primeur wines, which are kept in their barrels, until they go through the bottling process a few months later. Here again, members of the club are invited to come for the bottling. Both the purchase and the bottling gave opportunities for memorable visits to some famous chateaux.

The beauty of this club was the possibility to get together, outside formal business meetings, and enjoy fantastic wines at reasonable prices. The biggest joke at that time was the fact I did not drink wine, but I was participating in all the events.

For me, it was a sort of investment (the value of our bottles was always increasing) and a way to please my friends. We rented a cellar in downtown Toulouse to store all the bottles and each member used to help himself from his share.

Tim was a very active member of this club, with Don Foster from Emirates (both advising on the choice of the wines), with also the duo Henri de Sulzer and Georges Pons, from Airbus, who were taking care of the cellar and the accounting.

This wine club became the symbol of the close relations between the Airbus and Emirates teams, and became popular within the top executives of both companies, to the point that some of them, like John Leahy, decided to join years later.

We can say that the club strengthened the trust between the two teams and the two companies for a long period of time during which all the business dealings took place between the two teams, with minimum involvement of the top management in Toulouse.

Jean Pierson gave our team the full power to handle the relationship with Emirates, and he had even advised John Leahy, when he took over from Stuart Iddles in Autumn 1994, not to interfere in the relationship between the Middle East business unit and Emirates.

The wine club had its informal headquarters, in Tim's favourite restaurant, *Le Cantou*, located near Airbus and which is aligned on the main Toulouse airport runway. Eating in this restaurant was a pleasure because of its good food and its good wine, but it was a punishment for the ears because of the noise of the aircraft landing and taking off every few minutes (luckily, over time the aircraft were to become much quieter).

As of 1989, Airbus embarked on several cooperation projects with Emirates (particularly the joint venture for pilot training with C.A.E, the Canadian manufacturer of flight simulators) and helped Emirates in organising or participating in several international events (Golf Desert Classic tournament in 1989, rugby sevens tournament for several years, gala dinner of the Dubai Airshow.

These different sponsorships were a sort of offset obligation to assist Emirates in promoting its activities and in creating worldwide awareness about its innovations. Besides, Tim was also politely and lightly suggesting to Airbus to participate.

I had another friend in Emirates, Gary Chapman, who oversaw the other activities of Emirates Group besides the airline and who was a very convincing fellow, to make Airbus participate in financing all these events. To be fair, it was a good investment for Airbus because, as a result, Airbus had good public relations exposure during these different events as well.

These multiple partnerships allowed me to get close to several celebrities like the late Severino Ballesteros, the former Spanish golf champion with whom we had the opportunity to play a game as part of what is known as Pro-Am and which is an informal competition between professionals and the amateurs preceding the official competition, myself as a caddy for Michael Jones, an Airbus employee, Stuart Iddles and another person who was caddying for him.

I also had the opportunity to have lunch and dinner with him. I mention Ballesteros because we had a very funny conversation about back pain from which he was suffering, and I was surprised when he asked me if I could borrow him some pain-killer pills I was taking for the same problem, the night before the competition.

His only comment, after reading the leaflet describing the medicine was:

"Perfect, there is no forbidden content in it, thanks, I'll owe you my victory, if I win."

Later on, I understood the reasons behind his suffering, and I admired much more his courage and performance. His death at the age of 54 was a real shock and it saddened me. I liked the guy and the champion. For the first time, a non-Anglo Saxon golfer was able to reach the pinnacle of this elitist sport and become the leader for a while, followed by his colleague Jose Maria Olazabal.

I had other close encounters, on two different occasions, with Rod Stewart and Ian Dury from Ian Dury and the Blockheads. Ian Dury was singing at the end of a rugby sevens tournament, and we were staying at the same hotel. It happened that we came back to the hotel at the same time, and I was singing one of his famous songs, 'Reasons to be cheerful'.

He was walking behind me in the lobby of the hotel, and I heard him singing the same song as if he was trying to make a duo. I turned back to see him, and he smiled. I just told him that I enjoyed the show. He seemed to be tired, and I did not like to bother him. He died a few years later.

Following the 'Dead Rabbit' adventure, we embarked on several incremental orders for A300-600s and A310s, and we started working on the long-term fleet plan of Emirates. We had frank and transparent exchanges about the vision and the future plans of each company. During a visit by Jean Pierson, Sheikh Mohamed Bin Rashid Al Maktoum gave an idea of his vision of the future of the Emirates and the Dubai airport.

There was mention of more than 50 aircraft by the turn of the century and much more by 2010, with an airport of the size of the big European airports. Jean Pierson was very fond of Emirates but, nevertheless, he was somehow surprised by the ambitious plans described by Sheikh Mohamed.

Tim confirmed the ambitions of the Dubai leadership on several occasions after Pierson's visits.

In particular, he said something to me I'll never forget:

"Habib, listen, Emirates will be a big player in air transport, and Dubai will become a big hub. The leadership of Dubai is ambitious, and we know how to reach the targets."

I trusted him without any hesitation.

By the mid-90s, Airbus was contemplating the launch of a smaller-capacity version of the A330. We started working with Tim on what size and what range could suit the Emirates' needs. This was the first time we worked with Tim on aligning the strategies of the two companies.

This exercise was done in full transparency from both sides. Airbus presented its different projects (M15, M17, etc., M meaning minus, and the number represents how many frames were to be removed from the fuselage) and Emirates presented its expectations in terms of passenger capacity and maximum range.

Airbus was working secretly with a very limited number of airlines to evaluate the potential size of the market for a shrunk A330. For this purpose, Tim came to Toulouse for some confidential presentations and, having reached a preliminary understanding of the specifications of the aircraft, it was agreed to continue the discussions on the price and other contractual issues, immediately.

To ensure total confidentiality of these discussions, Stuart Iddles suggested the meetings to be held in London. Consequently, we travelled from Toulouse to London, separately. It looked like a James Bond movie. We met Tim, Stuart, and me, in a hotel room in London for close to one full day. Stuart left late afternoon, and Tim and I stayed overnight to continue the discussions.

This desire for secrecy was dictated by the fact that the A330 shrunk would be a real killer of the B767 for two simple reasons.

One is that the new A330 would have better performance, payload, comfort and economics than the B767, and the **other** was that Boeing was squeezing its own product with the B777-200.

Airbus had to choose the right timing to announce its new product and preempt a strong reaction from Boeing. Anyway, this was one of my important lessons in the strategy of communications. I understood that communicating seemed to be as important as doing. This was the opposite of what I had learnt in schools: action before blah blah blah.

For our A330 shrunk project with Emirates, I noticed, very quickly, a clear convergence of our points of view, and within a few months the A330-200, in its known configuration, was born, thanks to the Letter of Intent of Emirates for 16 aircraft signed early 1996. Some other airlines got a quick interest in the aircraft and signed contracts before Emirates, but the real launch customer for the A330-200 was Emirates (the final contract took a few more months to be signed).

The same scenario was repeated later for the A380 (Emirates was the first to commit and thereby was the actual launch customer for the programme, but another customer was to become the first operator). For the A340-500, Emirates was the launch customer and was also the first to take delivery.

The A330-200 exercise allowed me to know better Tim and learn how to challenge him intellectually. This was very useful when we later worked on the new first-class cabin concept with the private suites for the A340-500 and the A380.

Our first big adventure with Emirates and with Tim Clark was the negotiations of the A330-200 contract during the summer of 1996. We agreed on the aircraft, on its main features and on the number to be ordered (16). It was launched, following a Letter of Intent from Emirates and commitments (including firm purchase agreements) from other customers.

The launch of this aircraft was the result of joint efforts by all the sales regions, and I really enjoyed the spirit of cooperation and exchange which prevailed between the heads of the regions and their teams.

For different reasons, Emirates decided to start the contract negotiations by the end of June 1996. In order to ensure maximum efficiency, we jointly decided to book the main meeting room and some adjacent ones in the Jumeirah Beach Club hotel (actually the Four Seasons hotel in Dubai), with some adjacent rooms and to put the two negotiation teams there for as long as necessary, until the contract was signed.

For this purpose, everything was available on the spot, including good quality food, drinks, photocopiers, fax machines, etc. The only drawback was the fact that the meeting room had a fantastic view of the main swimming pool of the hotel. It was summer and people were swimming, while we were arguing and fighting for each word and each figure. One day, I had an open discussion with Tim about the way the negotiations were being conducted by Emirates.

His explanation was very clear:

"In the early days, Emirates bought Airbus aircraft without opening a real competition, and it was done by successive small purchases of two to three units. The Emirates team was new and did not have time to dig into all the details. Despite this, the experience with the A310s and the A300-600s can be considered successful, and Airbus did not let us down."

"But for the forthcoming fleet expansion, Emirates prefers to start with a clean sheet of paper. For this, the present contract must cover all aspects and serve as a basis for any other future contract with Airbus. In a nutshell, we are structuring the frame for our future relations with Airbus."

It was a clear and frank answer, which hurt a bit because I felt that Emirates was not giving full credit to Airbus for all the efforts made to help Emirates take off when nobody believed in this newcomer. But it had the merit to show that you must fight every day to keep the loyalty of your customers.

This being said, Tim's answer was a clear sign of the level of openness and transparency we had reached in our relations, and it was also good advice to Airbus to have a basic agreement which can serve for any future transaction. This was confirmed by the relatively short negotiations of the subsequent contracts signed with Emirates for the A340-500s and the A380s.

With Tim, we agreed on a draft agenda and a tentative timeline. My personal expectation was that the negotiations would last for three to five weeks. But the reality was different. The two teams were there, in this meeting room, for 97 days non-stop.

Some of the negotiators were there from the beginning to the end, like Henri De Sulzer and Gerry Sharp. Personally, I spent 73 days, including 80 hours non-stop (with intermittent sleeping breaks of two to three hours) which were imposed on all the negotiators to close the deal, before a given date. Tim paid also his share of negotiation time. After 45 days from the start, we were still far from any foreseeable closing date.

The contract clauses put in the bucket for further review were largely exceeding the closed ones. Consequently, I decided to go and see Tim to try to convince him to help me accelerate the negotiation. Knowing his philosophy of building a solid contract which could be the basis for any future transaction, I was not expecting miracles. But he acknowledged that the pace was slow and that we needed to move faster.

I understood from our discussion and without hearing him say the words, that the empowerment of some local executive deserves some time. That day, I discovered that Tim was a great diplomat and that he genuinely enjoyed building a solid company with the involvement of Emirati citizens in the highest possible positions. I did not push further.

But from that day on, we imposed on the teams a regular review of all the pending items with decisions taken on the spot. Thanks to this new approach we did not exceed the 97 days!

We finished the negotiations sleepless, at 5 am, but we signed the contract the same morning, before taking the afternoon Emirates flight to Paris or London.

After closing the A330-200 chapter, we immediately opened the A340-500 chapter. Here again, Emirates was a launch customer, but this time with a major innovation: the mini first-class suites. This was one of Tim's ideas. He had always shown a great interest in cabin configurations and interiors. He imposed some clever modifications to the seat and galley manufacturers to accommodate Emirates' special requests.

This time he imposed the concept of a closed space offered to each first-class passenger. We discussed the concept and he quickly admitted that there would be some regulatory challenges (for example, the ability for the cabin crew to see all the passengers during take-off and landing. This could, possibly, impose the need to install auxiliary cameras, how to unblock the door of the min-suite from the outside, amongst other things).

He also requested the installation of variable mood lighting in the ceiling, which was a real challenge for Airbus engineering.

Due to all these innovations, we spent months shuttling between Toulouse and Hamburg, discussing, designing, testing and fine-tuning these new features. Luckily the contract negotiations were relatively short during the summer of 1997 thanks to the existence of the A330-200 contract.

During this relatively long period of design of the A340-500 interior, I discovered Tim's skills as an architect and interior designer. We had frequent opportunities for serious discussions about matching colours, hiding some technical devices, while keeping them accessible, defining comfort space around the passenger, etc.

We got some interesting lessons from some professional designers and had some hilarious moments when we tested some unfinished equipment.

One of the consequences of working closely with Tim since the A330-200 project was the time we spent discussing other subjects, and in particular music and history. In this respect, we had a common attitude vis-à-vis what we called the *official history* as we had learnt at school. Following a heated debate in Dubai, with Tim and Ram Menen, the then-head of cargo at Emirates, Tim suggested I read a book called *Fingerprints of the Gods* written by Graham Hancock. At the same time, I was reading a book about the sphinx, explaining that the damage we see on its back is due to water erosion. Since it stopped raining on the Guizeh plateau more than nine thousand years ago, this erosion could have only taken place before around 8000 B.C. This contradicts the official age of the Sphinx which is officially dated to 2500 B.C.

Following this discussion, I became a loyal reader of Graham Hancock and Robert Beauval's books. And I attended, alone or jointly with Tim and Ram, some conferences about *real history* or *ancient studies* (to avoid provocative titles), where I met Hancock and Beauval and had the possibility to exchange directly with them. This common interest in the untold history of the world created a strong bond between Tim, Ram, and me.

Each time we met, we dedicated some time to discuss new developments, new publications and new theories. I must confess that this immersion in the world of Hancock and Beauval influenced a lot my vision of history and of the origin of humanity.

As from 1996, I started feeding Tim and the Emirates team with information about our A3XX, the future very large aircraft. Tim seemed very interested, and the frequency of exchanges increased to the point that my Emirates/A3XX paper file became quite substantial. Tim started commenting on some of the basic designs.

To give him a clear picture, I invited him to Toulouse in 1998, where I organised for him a meeting and a lunch with Jürgen Thomas, the Chief Engineer for the project. Tim and Jürgen enjoyed the discussions and had a very frank exchange of points of view. Jürgen took note of some key comments of Tim and about the maximum take-off weight and the range.

This exercise continued for a few months, until 1999, when Airbus organised a specification review of the A3XX at the Hilton Suffren in Paris, for the potential launch customers. Before attending the meeting, I had a coordination review with Tim. We went to the conference as a single team.

Following a very precise review, Tim made some suggestions like fine-tuning the shaping of the forward main stairs, adding another pallet in the cargo hold, repositioning some elements of the rear upper deck galley, increasing the water quantity in the tanks to cope with the installation of showers in first class and some additional minor items.

One can see the difference between airlines willing to bring another standard of service with the A3XX, and airlines considering the A3XX as a simple double fuselage which would increase the capacity but could be furnished as usual (we saw some airlines using later what became the A380 as a double B777 with the same seats and the same service).

Following this review in Paris, I understood that Tim was taking Emirates to another league, where there are only a few competitors and where they will, for sure, be number one.

I remember his comment during a dinner in Paris:

"This aircraft is an exceptional real estate. We must optimise it and get the best out of it."

He insisted that there was no contradiction between a high number of passengers and a high quality of service.

After the Paris gathering, we entered into detailed discussions with Emirates about a potential commitment of Emirates for the aircraft. In this respect, Tim started giving an idea about the price bracket. Emirates had already committed to the B777-300 and knew its price. Tim was applying a sort of rule of three.

Nevertheless, we had a serious discussion about the maximum acceptable price level, and we came to a sort of compromise. A few weeks later, we had an internal discussion within Airbus with the CEO to define a range of launch customers' prices. I suggested the figures discussed with Tim, and at the end we came close to them.

By the end of March 2000, Airbus and Emirates started serious negotiations about a Letter of Intent for the A380. The only point of disagreement was the number of units. Emirates initially suggested eight firms and five options. We asked for more. Finally, we agreed on 13 firms and 5 options.

On 13 April 2000, the first Letter of Intent for the A380 was signed in Dubai at the Crown Plaza Hotel. Tim and I were the happiest persons in the ballroom where the signature took place. This happiness was motivated by the feeling of *mission accomplished*. We were anxious to see the A380 launched, and the first step was achieved.

A few months later, at the Farnborough Airshow, a formal Memorandum of Understanding was signed for 22 A380 passenger aircraft and two freighters, at the same time as Air France for a dozen aircraft. It was not acceptable to launch the aircraft without one customer from the four partner countries (France, Germany, UK, and Spain).

This led to the formal launch of the A380 programme, hereby putting an end to the monopoly Boeing had enjoyed for decades with the 'Queen of the Skies', the B747, which was Boeing's 'hen with the golden eggs'. This also enabled Airbus to offer a complete product range, from the small single-aisle aircraft to the very large, very long-range ones, hereby competing head to head with Boeing—the only other global airliner manufacturer in the world—in all market segments.

The firm purchase agreement (final contract) with Emirates was signed during the Dubai Airshow in November 2001, for the same quantities, just a few weeks after 11 September.

I arrived in Dubai, on 10 September to kick off the negotiations of the A380 contract and to start working on the new service company (TASC) Airbus had decided to set up in Dubai jointly with Emirates (as explained previously). The next day, I saw live the second aircraft hitting the second tower of the World Trade Centre in New York, while I was standing in the lobby of the Jumeirah Beach Club Hotel.

I just came back, late afternoon, from my daily run and was sweating like a steam machine when Doris, the general manager of the hotel, called me inside the lobby and pushed me in front of the TV set, telling me:

"Look at what is happening in New York! The Twin Towers are destroyed."

CNN was broadcasting in a loop the crash of the second aircraft which had taken place just a few minutes earlier. I could not grasp what was happening and I was not focusing my attention, because I was feeling very cold, due to the air conditioning. Doris, who was keen to share with me the event, brought me a huge towel and started explaining to me the sequence of events.

When I understood the full picture, my first conclusion was:

"Game over."

I called Tim and we sat together, two days later, for a working lunch at my hotel.

After a complete review of the political implications of this attack in the US, in Europe, and in the Middle East, and particularly in the aviation field, we discussed the consequences for Dubai and the UAE. Tim, who was himself shocked by what had happened and concerned about the future, was nevertheless very optimistic while taking into consideration the seriousness of the event.

He told me without hesitation that the US would retaliate one way or another, maybe by choosing the wrong culprit. For him, the world would be in a wait-and-see situation for a few days or weeks, but things would eventually come back to normal. Air transport will see more stringent security measures, but above all, Dubai would continue on the same track, keeping the same objectives, as already explained to Pierson, more than a decade earlier.

The negotiations of the A380 contract would start as planned, with the target to sign at the Dubai Airshow in the fall of 2001. I was agreeing with him on the fact that the US would retaliate and that air travel would become more complicated at airports, for the passengers, but I was more pessimistic about the pace with which we would return to business as usual.

I had an additional concern, due to my origins, citizens of Arab countries would suffer when travelling to the US. This was quickly confirmed by some friends who dared travelling to the US, including Jean Pierson who saw his visa rejected because he was born in Tunisia. I remember him having cancelled his trip and the conference he was supposed to give.

It took some top US executives and even some senators to intervene and convince him to reschedule his trip and also to convince the US embassy in Paris that there was a mistake. Personally, I decided not to travel anymore to the US to avoid the problems some of my Arab friends went through.

I have not visited the States, since early 2001.

As for the attitude of Dubai and Emirates, Tim was right. I always saw this consistency in following the vision of the leadership and in keeping the set objectives, despite many hurdles encountered on their track. I remember the first Gulf War (Iran/Iraq), the second Gulf War (invasion of Kuwait), the dispute with Iran about the islands, and a few less important hick-ups.

The answer of Sheikh Ahmed, Maurice Flanagan, or Tim was always the same: *Keep on running.*

The bold attitude of Emirates generated for me continuous updates to Airbus top management about the rationale of the airline's moves.

I remember that the late Jean-Luc Lagardère, who was chairman of the board of EADS, our mother company since 2000, and who was planning to attend the A380 contract signing ceremony at the Dubai Airshow, requested a comprehensive presentation about Emirates' strategy and objectives and how confident we could be in their ability to achieve their plans.

This was the third or fourth exercise I did about Emirates since their decision to purchase 16 A330-200.

Finally, the contract was signed as planned, at the same time as a Letter of Intent to set up the joint venture TASC. Lagardère was present.

We had had dinner together the evening before the signature, during which he made a few nice comments, and after the signature, he came to shake my hand with a big smile and a warm:

"Bravo, merci beaucoup."

During the whole airshow, no word was exchanged about my previously mentioned resignation, neither with Lagardère, nor Forgeard, or Leahy. It was also the last time I saw Mr Lagardère alive. He died in 2003.

As for the marketing studies about the Emirates' objectives and possibilities, each time I had a question from my management concerning Emirates and their intentions, I asked Tim to give me some ammunition, and I could count on my marketing team to help me put together all the different elements to substantiate their claims and satisfy my management.

I had the impression to pass an exam each time Emirates made a bold decision to buy a big chunk of Airbus aircraft. But, at the same time, I was somehow proud that the marketing rationale I had introduced many years ago was properly used. And to be fair, both Louis Gallois and Tom Enders also trusted me and did not ask me to provide a marketing rationale concerning any further Emirates purchase.

Tim knew about these challenges (of convincing the Airbus management) and was always available to help with the necessary data. It must be underlined that such marketing rationale was also useful for the preparation of the requests to be submitted to the European Credit Agencies (ECAs) to get their guarantees for the bank loans needed to finance the purchased aircraft.

There were other cases where Tim and I worked together to prepare important documents, either to convince the civil aviation authorities of some European countries to give more traffic rights to Emirates or to convince the European Union that Emirates is not a subsidised airline with below-market wages and unacceptable working conditions.

For the traffic rights, we developed a routine. Our teams prepared the necessary documents, we reviewed them with Tim, then we requested an appointment with the relevant minister, and I went with Sheikh Ahmed to meet him. Sheikh Ahmed had a strong charisma, and he always succeeded in convincing his counterpart of the validity of the request.

But there is one memorable occasion worth mentioning. It was the first meeting between Mr Jean-Claude Gayssot, a member of the communist party who, at the time, was the French transport minister in the Jospin government, and Sheikh Ahmed.

Both were heavy smokers, and at a certain point, I saw Mr Gayssot offering one of his cigarettes to Sheikh Ahmed, who accepted it with pleasure. And I saw Sheikh Ahmed lighting the cigarette of Mr Gayssot with his own lighter. Despite the barrier of language, the two were communicating somehow.

And the assistant of Mr Gayssot and myself, besides translating some words from time to time, we were just sidelined. Sheikh Ahmed enjoyed the meeting and the warmth of Mr Gayssot. It seemed there was a special bond between smokers. When we left Mr Gayssot's office, we knew that the additional frequency would be granted.

A few weeks after the contract was signed, with the big echoes it had in the aviation world, Tim called me for a business review. First, he announced to me the bad news: Emirates cannot be part of the planned joint venture with Airbus, despite the signature of a Memorandum of Understanding during the airshow. The reason? Some advice from their legal team to avoid any conflict of interest.

So be it. TASC would be a 100% Airbus subsidiary. I did not grasp the full consequences of this decision, but Emirates had provided Airbus with an elegant solution to solve a problem which will be further explained below.

Then, Tim announced the good news: the updated fleet plan showed a need for several additional A380s, but the delivery dates must be acceptable.

It was my turn to make some announcements to Tim. I confirmed to him that I had resigned from my position of Senior Vice President of Sales as of June 2001 and that the plan was to proceed with the setup of TASC and to run it from Dubai.

Before I finished my explanation, he interrupted me:

"But who will be in charge of Emirates?"

We have already discussed this matter with the CEO, Noël Forgeard, and the Chief Commercial Officer, John Leahy. A possible solution was envisaged, but it needed a legal opinion. I promised to give him an answer very soon.

A lengthy discussion followed about the reasons for my resignation, which were explained above. Tim reacted by saying that the results were there, and Airbus had become number one.

I said: "Yes, Airbus is number one in the market, but the CEO and the COO consider that I am too emotional (somehow Oriental)." They forget that emotions are necessary in any relationship, especially if they are well-controlled. Emotions are the basis for personal relationships, which are so important in business. One of the reasons of our successful and trustful relationship with Emirates was that both sides could express their emotions, and I believe Tim would confirm that.

Early in 2002, I moved to Dubai, to be based there permanently, and with the help of a small team from Airbus, particularly Christophe Fanjat, the TASC business plan, statutes, and organisation were ready.

We decided to establish TASC at the recently created Dubai Airport Free Zone Authority business office compound (DAFZA), next to the airport, and leased some offices there.

At the same time, we confirmed to Tim how Airbus would continue to manage the relations with Emirates. The solution was that TASC would be charged by Airbus to ensure the coordination of all relations with Emirates through an arm's-length agreement.

The concept was validated with Sheikh Ahmed, Maurice Flanagan, and Tim Clark before it was implemented. It must be said that all three had insisted vis-à-vis Airbus that I continue to be the Airbus interface with Emirates.

In July 2002, TASC started its operations, and I continued to take care of the Emirates account. Between Tim and me, it was business as usual. This solution would not have been applicable if Emirates had remained a partner in TASC. Bad news can generate new and positive opportunities.

My presence in Dubai, as from 2002, allowed me to have more frequent contacts and discussions with Tim. One can say that we developed a real friendship. The strength of our relation was demonstrated in March 2004 when I had a serious health problem and was, for several days, at the American Hospital in Dubai. Tim came to my room and spent long moments with me. I could see that he was concerned about my health.

Tim suffered also from a similar problem at a later stage, and I went to see him at his house. From that date on, we agreed to have our yearly check-ups together at the same place, with Professor Jean Marco, who treated me initially in Toulouse before moving to Monaco. Each year, Tim, John Leahy, and I used to travel, for two days to Monaco to get our check-ups.

It was an opportunity to be together and enjoy the nice weather and the good food. After the checks and the serious advice for a good diet, all of us, including Professor Marco, used to have, together, a heavy lunch or dinner, which would contradict all the recommendations.

I stopped this ritual after the retirement of Professor Marco. John Leahy did the same, but Tim continued to go alone.

In 2005, while TASC was celebrating some notable successes, Airbus decided, as explained previously, to develop the services activity internally through the Customers Services Directorate, reducing de facto TASC's portfolio of business and putting some serious doubts about the future of the company.

During the same year, Airbus also decided to launch the first version of the A350 (an upgraded A330 with new engines). This was the initial Airbus response to the launch of the B787 by Boeing (in my view, and in the view of many others within Airbus, it was a hasty decision)

After the official presentation of the A380, or 'Reveal,' in Toulouse, on 18 January 2005, I was called to a briefing on the A350 and tasked with convincing Tim and Emirates to buy the aircraft to replace their A330s. I was not convinced by Airbus' decision, but I went to see Tim and presented to him this initial *A350* concept. It just took me five minutes to get the confirmation that he would not buy it.

He told me bluntly:

"Habib, it is a no-brainer; this aircraft cannot compete against the B787, and I don't see how it could fit in our fleet development strategy. I advise Airbus to review their copy."

Despite the glamorous A380 Reveal ceremony in early 2005, Tim was not in a good mood vis-à-vis Airbus anymore because there were rumours about some problems with the A380 programme (a slight delay and some small overweight). It seems that Tim heard the rumours from some other customers who were attending the ceremony.

I was surprised by the importance he gave to these rumours because we had regular briefings, and there was nothing hidden, neither for the overweight nor for a possible short delay. In these circumstances, and in the middle of continuous briefings about the progress of the A380 up to the first flight planned in April, I tried to continue marketing the A350 within Emirates.

The only tangible result was a polite:

"Let me see," from Tim.

The first flight of the A380, which took place on 27 April 2005, in Toulouse, brought back some positive feelings to everybody on both sides, between Emirates and Airbus, and towards Tim. Encouraged by the positive mood and despite my advice, Noël Forgeard and John Leahy insisted in making themselves a direct push with Sheikh Ahmed and Tim and decided to travel to Dubai in early May, together with Christian Scherer.

They had also the duty to update Emirates on the progress of the A380 and announce the, by now, expected delay. They arrived in Dubai two days after I was admitted into hospital in Dubai for severe nosebleeding due to the combination of an excess of red cells and the fact I was taking blood thinner medicine to protect my heart.

The group went to see Sheikh Ahmed and Tim without me and came later to brief me at the hospital. It was a surrealist scene, where the three visitors were seated around my bed, debriefing me in detail, and trying to understand what I was saying, because of my nasal voice due to the blocking of my nose. In addition, my wife told me that the smell of the blocking nasal wicks was somewhat bad, hence the relative distance between the visitors and myself.

This made it difficult for Noël, John, and Christian to understand clearly what I was saying. From this unscheduled visit, I understood that they were desperate to get commitments from key airlines to launch the A350 programme. Tim, who called me to check on my health, was surprised by this visit and did not show any change of attitude towards the proposed aircraft.

His only noticeable comment was:

"Tell them to have a serious review of the real needs of the customers and to come up with a true solution."

This was a clear call for a change of the aircraft specifications.

Following this exchange with Tim, I wrote a comprehensive report to Toulouse. I thought the management would review the situation, but nothing happened.

Meanwhile, it must be mentioned that, on 4 May, Airbus announced a six-month delay of the first A380 and another two to six months delay for all the customers. This delay was not appreciated by Tim, who started showing a clear dissatisfaction vis-à-vis Airbus. But since he was kind of expecting this delay, he did not blow up, neither against the Airbus delegation nor against me.

After this announcement and after I recovered from my bleeding nose problem, I had to fly to London for one of the funniest business meetings in my whole life. Trying to make another push, John asked me to organise a meeting with Tim in London. To ensure a relaxed atmosphere, we booked a table at the *Waterside* restaurant on the Thames. Before lunch, we had the brilliant idea to have the aperitif while sailing on a small boat on the river.

Due to our lack of experience in manning properly a boat and to the quality of the aperitif, John, who was steering the boat, was close to crashing into a rowing team which was in an intensive training session. The coach of the rowers expressed his anger. I tried to take over the helm from John, but finally, Tim took it.

The easiest option was to continue straight and dock at the spot we departed from, but we saw a few strong guys standing there, watching us with very bad eyes, expressing strong anger. Tim carefully stirred away the boat, and we docked a little further down (or up?) the river, closer to the restaurant. It was a real strategic retreat. Needless to mention that we were the focus of the majority of the restaurant's clients, who could have a sight of the docking area.

Eventually, we settled at our reserved table, but the combined effect of the aperitif and the good wine provoked louder voices, mainly from John. I believe half of the restaurant had a clue about the content of our discussion.

They must have insulted us multiple times because people struggle to get a table at this restaurant, and they go there to enjoy its quietness, its top-quality food, and the beauty of the landscape. We must have been seen as impolite and misbehaving intruders. Tim was half convinced by John's arguments, and the A350 saga was put on hold for a while.

Another important development took place in the spring of 2005. On 26 May 2005, the appointment of Noël Forgeard as co-CEO of EADS, jointly with Tom Enders for the German side, was eventually confirmed and formalised by the EADS Board (since 2000, EADS was the Airbus mother company). The German Gustav Humbert succeeded Noël as CEO of Airbus.

Noël Forgeard had at last fulfilled his dream of becoming the EADS CEO. I was happy for him, but at the same time, I was somewhat anxious because the simultaneous change of management, both at Airbus and at EADS, took place in the middle of the announcement of the A380 delays and of the slight overweight of the aircraft. There was also the *false* start of the A350.

In addition, I knew that the appointments were taking place in the middle of some tensions with the German partner. I was expecting a serious recovery plan, including all aspects, with timelines, to answer customers' interrogations.

Gustav Humbert inherited a very difficult situation, with a lot of simultaneous problems, and I had no idea what the priorities of Noël Forgeard were.

I spent the whole summers of 2005 organising several briefings on the A380, putting aside the A350. But I created a small task force comprising two of our best marketing people, Grainne de Courcey and Peter Barnes, to help me write a full report to Gustav Humbert about how the A350 must be so as to respond to the real customer needs. This report was given to Gustav Humbert on 20 December 2005.

The report defined the main characteristics of the aircraft and gave the diameter of the fuselage, which was the one eventually retained for the new version, dubbed the A350 XWB. Gustav called me between Christmas and New Year to tell me that he liked the content and that he would discuss it with the people in charge (Olivier Andries and Philippe Jarry).

I told Tim about this report and gave him an overview of the content. He did not give the matter a lot of attention.

It must be underlined that the initial A350 was officially launched in October 2005, with 140 firm orders.

From December 2005, and due to the combination of the already announced A380 delay, and other technical problems, and the continuous push for a—to him—not satisfactory A350, Tim started showing a sharp change in attitude vis-à-vis Airbus.

He became more critical and lost the remaining confidence he still had in the company. He became suspicious that there were further problems with the A380 and totally rejected the proposed A350.

His confidence had started melting years before with the saga of the close to three tonnes of overweight of the A340-500, discovered in late 2000. I remember there was a sort of denial from Airbus for a few weeks before it clearly acknowledged the overweight. Tim thought it was either an attempt to hide the facts, or simply that the top management ignored the reality. For him, in both cases, it was an unforgivable sin.

The bare reality was that the wings suffered an overweight. It took some time to discover the problem, and a weight-saving plan was put in place, but it did not achieve the full target. Unfortunately, it was necessary to weigh the wings before defining the exact amount of overweight, which explains the hesitation of the top management to first acknowledge there was an overweight problem and second, to give an exact figure.

Tim was unhappy with this situation and used a specific expression to describe the problem: *wings of concrete.*

It was the first time I discovered that Tim could get very upset, be bloody, and could say harsh words. I lived through similar situations in the following years until my retirement.

Despite many positive developments on the A380 programme (first flight, three aircraft flying, first landings at major airports all over the world, performance meeting expectations), Tim continued to be critical. There were frequent technical reviews in Dubai to alleviate his fears, culminating with serious clashes during the first quarter of 2006, especially after the news that Daimler-Chrysler and Lagardère were progressively withdrawing from the capital of EADS.

Tim was sure there was going to be a bigger delay than what was announced and that either Airbus was hiding it or unable to apprehend the extent of the problem. For him, the move of the big shareholders was not a good omen.

Unfortunately, the announcement of a second delay of six to seven months in June 2006, the replacement of Noël Forgeard by Louis Gallois, and of Gustav Humbert by Christian Streiff, confirmed Tim's fears and exasperated the tension between Emirates and Airbus. We could say that we reached level zero in confidence.

The following events—with the inside trading accusations against many Airbus executives who sold their stock options before the announcement of the third delay, the effective announcement of the bigger delay (18 months in total), and the departure of Christian Streiff after only three months in the job—made things more complicated with Emirates, particularly with Tim and Adel Redha, the Executive Vice President of Technical and Operations at Emirates.

Luckily, the arrival of Louis Gallois as interim CEO of Airbus in October 2006, while keeping his position as co-CEO of EADS, brought back a more balanced relationship with Airbus.

My frustration during all this troubled period was that, despite the fact I had kept a friendly relationship with all the Emirates management, their anger against Toulouse made them insensitive to my arguments. For them, I was still their friend, they trusted me, but they believed that Toulouse was *intoxicating everybody, including its own people.*

A parallel development took place during the first and second quarters of 2006. Following my report on the A350 and other exchanges with Gustav Humbert, he asked me to come back to Airbus.

After a first refusal, he came back with an innovative scheme, whereby I would stay in Dubai to manage a new subsidiary of Airbus called Airbus Middle East (AME), regrouping sales, marketing, contracts, communications, and customer support, including a spare centre and pilot training organisation, and covering the MENA region (excluding Tunisia and including Iran, Pakistan, Afghanistan, Sri Lanka, Ethiopia, Eritrea, Djibouti, and Senegal).

Due to our proximity as neighbours in Toulouse and a friendly relationship for several years, Gustav and I were able to discuss openly and in full transparency several subjects. I believe he listened to me carefully.

After the harsh comments from Steven Udvar Hazy, president and founder of ILFC (one of the two biggest leasing companies in the world at that time), in April 2006, concerning the need to review completely the specifications of the A350, Gustav called me to have a coffee at his home to discuss several subjects related to Airbus.

He came up with three conclusions: a full audit of the A380 programme in order to have a precise idea about the global situation and define the exact extent of the delay; complete the review of the A350 specifications and come back to the market with the new concept; and finally, start a restructuring of the company in order to avoid the shortfalls experienced in the past few years.

He confirmed to me that the report I had produced jointly with my team about the A350 was met with a certain resistance from some key players within Airbus, but the number of comments received from the customers strengthened the validity of the report.

In this respect, I knew that, as of late 2005, there were new iterations to redefine a more attractive and efficient A350, but there were still hopes from certain people that the original version would do the job.

AME started its operations in June 2006, in the middle of the big turmoil related to the A380 delays. The new A350 XWB was eventually announced by Christian Streiff during Farnborough Airshow in July of that year.

From a certain perspective, the start of AME coincided with a huge clarification of the Airbus landscape and brought back some balance in the relations with Emirates and Tim Clark. A settlement was signed concerning the new delivery dates of the A380, and we had to focus on the preparation of the first delivery to Emirates.

Regrettably, despite being the first customer of the aircraft, Emirates would receive its aircraft after Singapore Airlines because Rolls Royce was the first certified engine, and Emirates had chosen the Engine Alliance (an association between General Electric and Pratt & Whitney) one.

Taking advantage of the more positive mood that was prevailing as off mid-2006, we succeeded in marketing the new A350 XWB with Emirates and signing a contract for 70 A350 XWB (50 A350-900 and 20 A350-1000) during the Dubai Airshow in the last quarter of 2007. The relative honeymoon continued until the delivery of the first A380 in July 2008, but this was to be the last happy moment for a long time.

Nevertheless, the A380 delay necessitated a crash action with Emirates to avoid tough settlements and even possible cancellations. This covered the years 2006, 2007, and a large part of 2008. But despite the Damocles sword on our heads, the situation could be described as paradoxical because it was an explosive mixture of cancellation risks and still some desire to buy more aircraft from Airbus.

In a nutshell, I spent the whole period between late 2008 and late 2014 riding a roller coaster with Emirates and Tim. This situation impacted my personal life and the life of Airbus and had a huge effect on certain Airbus aircraft types.

July 2008 was a very busy month for Airbus, Emirates, Tim, and me. A preparatory meeting for the upcoming delivery of the A380 took place just after the 14 July, or *Bastille Day,* celebration in France. Tim and a group of Emirates public relations and communication team came to Hamburg, to fine-tune all aspects of the delivery ceremony with their Airbus counterparts.

Following this meeting, Tim and I met in London on 23 July to inaugurate the biggest and the heaviest A380 model with Emirates colours, installed on the previously known Heathrow Concorde roundabout, which hereby became the Emirates roundabout. This model, which weighs 45 tonnes, is 24 metres long and has a wingspan of 26 metres. It is a one-third scale of the real aircraft and is comparable in size to a business jet.

The British press was somehow critical of the shyness of their national carrier compared with Emirates, and this for several reasons. The location had been owned by British Airways, which had been displaying on it the famous Concorde—the airline's flagship and that of the British aircraft industry (jointly with the French)—for many years. But then British Airways sold the location to Emirates, which replaced the Concorde with its own flagship, the A380.

As one newspaper portrayed it:

"Emirates is playing on British Airways' turf."

This prediction was confirmed in the following years, with Emirates operating close to 10 daily A380 flights out of Heathrow and the other London airports, as well as close to 16 daily flights out of the UK.

As Tim said during the inaugural:

"While the previous Concorde model represented the past, our A380 represents the future, and it is a future with cleaner and quieter aircraft."

This appropriation of the famous Heathrow roundabout and the inaugural of the A380 model represents the highest-profile advertising for the A380 and for Emirates—one week before the first delivery of the aircraft to the Dubai-based airline. This was a masterpiece of communications, as good as the previous one by Pan Am with the B707 and the Beatles.

The only difference is that the new technology allows more efficient communication and coverage than what was possible in 1964. Tim was not in good physical shape during the inaugural, as he had a serious injury in his upper lip, and he was obliged to ask for Photoshop to remove the scar from the photo shots.

I also remember that one journalist had asked me how much Airbus had paid for the model, and I answered him that in France, there is a say:

"When you love, you don't count."

He replied:

"I understand Airbus loves Emirates."

And I answered:

"We are the parents of a cherished child."

I don't recall if that short interview was published. I was not supposed to deliver any speech or to make any declaration. Tim was the guest star, and it was an Emirates show. I was provoked on the sidelines by one journalist, and I believe my answer did not satisfy his expectations.

A week later, and more precisely on 28 July 2008, Emirates took delivery of its first A380 during a memorable ceremony in Hamburg, in the presence of Sheikh Ahmed, Louis Gallois, Tom Enders, and all the top executives of both companies, plus the two teams involved in the project.

It was an emotional ceremony, and I was close to crying. For me, it was the pinnacle of a long and exhausting adventure. For over 10 years, I had been pushing like mad to make this project materialise Tim Clark had been a very solid and loyal partner who had helped me in this endeavour.

Together, we worked hard to make it happen, despite some hesitations from the Airbus ownership and top management, and some reluctance from the usual big airlines which were dominating the market—trying to protect their turf but without the vision Tim had.

Tim was also emotional during the delivery ceremony. We looked at each other, during the speeches and the exchange of presents without saying a word. We simply exchanged smiles.

There was no need to mention our names—everybody in the delivery hall knew who did it.

At the end of the speeches, Sheikh Ahmed, Louis Gallois, and Tom came to see us both and congratulated us.

Our baby was born and was starting its life. And still today, I have a very special feeling when I board an Emirates A380.

After a highly publicised A380 entry into service with Emirates and the buzz created by the bar and the two showers on the upper deck, which made both Airbus and Emirates teams fly high for a certain period of time, we were forced to fly low and even land due to the bare reality of the daily operations (some unscheduled snags, followed by the finding of cracks in some parts of the wings).

These technical problems, which were a serious source of concern to Emirates and particularly to the Engineering Department of Emirates and its boss, Mr Adel Redha, provoked a dramatic degradation of the relations between the two companies.

I received dozens of late-night or early-morning phone calls from a very angry Tim who, I could notice, while trying to be polite vis-à-vis me, did not hesitate to criticise and say harsh words about a wide range of Airbus executives. I was a sort of shock absorber or anger burst damper. I spent months, not to say years, having to cope with such a crisis.

I still could have, from time to time, a friendly get-together with Tim, but these opportunities became rarer. On the opposite side, the number of harsh discussions or negotiations about technical problems became more frequent.

Airbus and Emirates signed some settlement agreement to cover the initial technical snags and, in particular, the wing rib cracks.

During the same period, Emirates became fond of the B777 and decided to cancel its plan to operate a fleet of A340-500 and A340-600. But at the same time, they wanted to increase the size of their A380 fleet. As of 2008/2009, Emirates focused solely on the A380 from Airbus but was impressed by all the wide bodies from Boeing. I started getting nervous about the viability of the A340-500/600 and the A350 contracts.

Another development took place in the aviation arena in Dubai—the creation of Dubai Aerospace Enterprise, or DAE, which was supposed to be the aircraft leasing arm of Dubai as well as an incubator for several aviation activities (training, manufacturing, servicing, etc.).

I was personally picked as a potential candidate for the position of managing director of the company and had one interview with the head-hunter and another with one member of the board (Mr Mohamed Al Zaarouni, CEO of the Airport Free Zone Authority, DAFZA). In the end, they chose an American citizen to run the company.

This new MD was very ambitious—may be too ambitious—and decided to purchase more than 200 aircraft, split 50/50 between Airbus and Boeing, while also embarking on many simultaneous costly projects. Unfortunately, for several reasons, and particularly the 2008 financial crisis, his ambitions for DAE were broken.

By the end of 2009, the company decided to downsize its activities and reshuffle the management. Consequently, the caretakers appointed by Sheikh Ahmed—the new Managing Director of DAE, Khalifa Al Daboos, and a small committee comprising one lawyer, one financial expert, and one member of Investment Corporation Dubai, the owner of Emirates and DAE—contacted me to discuss the potential cancellation of the 100-aircraft order. The same was done with Boeing.

While I started analysing the request of DAE, Tim expressed some concerns with a previous order for 12 A340-600. This was a clear consequence of Emirates' desire to use the B777 as its single tool to operate the long-haul market and of the dissatisfaction Tim had had with the A340-500 operations (mainly overweight and higher fuel burn).

The first confirmation of my fears came in late 2009 when Tim told me that Emirates wanted to cancel the A340-600 contract. I did not need to ask why. The answer was very obvious—an admiration for the B777. It took both teams close to six months to find an acceptable solution consisting in cancelling 12 A340-600 and all the DAE contracts against buying 32 A380s.

An agreement was signed to this effect during the Berlin Airshow in June 2010. This seemed to be a success for Airbus, with a total of 90 A380 sold cumulatively so far to Emirates. But the sad reality was that this was also the last nail in the coffin of the A340 family.

Berlin was an opportunity for Emirates, with Airbus' help, to put in place a very intelligent communication campaign vis-à-vis the German authorities and the German public, and also vis-à-vis the European authorities (the European Union was conducting an audit of the Gulf carriers to check if they use social dumping with their employees and if they were subsidised by their governments, in order to decide whether to grant additional traffic rights to European cities).

The signature took place in the presence of Chancellor Angela Merkel, under the wing of an Emirates A380, which was displayed simultaneously with a Lufthansa A380. The Emirates A380 got more attention from the Chancellor, which infuriated Lufthansa and pleased Sheikh Ahmed, Maurice Flanagan, and Tim.

Tom Enders, who was the Airbus CEO and the master of ceremony at the Berlin signature ceremony, was very proud of the exposure both Airbus and Emirates got during the Airshow. He introduced me to the Chancellor and explained to her my role in the company. Mrs Merkel took a few minutes to exchange some words with me in English, asking me what Airbus' market share in the Middle East was and concluding with a double *Danke, Danke*.

Mrs Merkel was very attentive to all our explanations and spent a comparatively long moment with both teams. Tim was very satisfied with the whole ceremony and had also a few minutes with the Chancellor.

This Berlin Airshow was preceded by an IATA Annual General Meeting (in short, AGM), with Airbus hosting the gala dinner and inviting the astronaut Neil Armstrong as a guest of honour. This was my second opportunity to meet personally one astronaut who walked on the moon, after Conrad.

The IATA AGM gala dinners used to be hosted alternatively—one year by Airbus and the other year by Boeing—while the other manufacturers hosted the breakfasts and the lunches. Tom Enders wished to mark the 2010 event with some very specific landmarks.

First, he chose an iconic location, which was the Tempelhof airport downtown Berlin, with all the memories attached to this emblematic airport and especially the air bridge organised by the Allied forces to break the isolation of West Berlin following the closure of its access by the Soviets with a wall built in August 1961.

Second, he brought a Junker 52, which was one of the most used civil aircraft before WWII and which had been the backbone of the Lufthansa fleet for many years, and one of the DC-3s, which were used for the Berlin air bridge, to be exhibited on the tarmac.

And third, he invited Neil Armstrong as a guest of honour. I believe he reached his target by giving a visual summary of aviation development from the 30s to date and by showing the different roles aviation can play in the lives of nations.

The presence of the Junker 52 reminded me of another flight display of this iconic aircraft, one year later, at the 2011 Le Bourget Airshow, to celebrate some anniversary. President Sarkozy was present, and Louis Gallois, the EADS CEO, was seated close to him.

The JU52, which was part of a group of vintage aircraft, was painted in the colours of the Luftwaffe, with all the details and with the name of EADS painted on one door (the aircraft is the property of EADS through its German part, as well as a Messerschmitt 262, which also flew during the 2010 ILA Berlin Airshow).

The JU52 was part of the Legion Condor fleet, which was very active during the Spanish Civil War, helping the army of Francisco Franco to fight against the Republicans.

President Sarkozy became blemished when he saw the JU52 and expressed his discontent to Louis Gallois.

During the 2010 ILA Airshow, the JU52 was painted in a plain dark blue colour, as was the case for the airliner versions. Therefore, there was no risk of upsetting any officials.

I believe this IATA AGM dinner remains for me the most remembered and appreciated event up till now.

As the protocol stipulates, the seating configuration must put the Director General of IATA (at the time Giovanni Bisignani) and the current President of IATA (the head of Lufthansa for this session) next to the guest of honour, plus the president of Airbus and his top guests, who happened to be Sheikh Ahmed and Tim Clark.

Of course, I had to be close to them, hence the opportunity for me to be seated among the ten people surrounding Neil Armstrong. This privileged position allowed me to participate in the lengthy discussion, which took place during the dinner of close to three hours.

For some reason, one guest was late and his seat, which was close to the centre of the table, was empty. Tom tried to know if the guest was coming or not in order to fill the gap and rearrange the seating as needed.

He got two contradictory answers:

"Yes, he is coming," and, "No, he is not coming."

This was an opportunity to start one of the most interesting and out-of-context discussions and to give me the opportunity to move to this seat and get closer to Neil Armstrong.

Tom reacted with some frustration:

"He is in, and he is out."

And somebody made a funny and uncommon comment:

"This is the paradox of the Schrödinger cat, dead and alive at the same time."

The whole group around Armstrong exploded in big laughter, and one guest said:

"Add Heisenberg's uncertainty principle and Dirac principle, and we will be completely lost."

At this point, I reacted by saying that he forgot the quantum entanglement, which will put us in a metaphysical world.

Neil Armstrong, who was listening up to now with a smile, intervened by saying:

"I wonder if quantum mechanics is not at the boundary between physics and metaphysics."

One other guest added a comment about the apparent contradiction between relativity and quantum mechanics and how this makes things more complicated to understand. I was surprised by the level of knowledge these people around me had about very complicated scientific matters, and I was puzzled by Neil Armstrong's comment.

What I did not expect was that this discussion would generate a lot of passionate exchanges among the participants and make them forget that the starter mentioned on the menu was already served. Luckily, the guest who was the cause of the discussion did not show up, and I could stay in that seat.

I thought the discussion was over when the guests started eating, but Neil Armstrong came up with a new comment:

"There are a lot of approaches to unify relativity and quantum mechanics, and scientists like Stephen Hawking have already explained some cosmological phenomena by using the two theories."

The debate saw a rebound, and this time we followed Armstrong's comment. The discussion focused on the Big Bang, the evolution of the universe, and the question of the existence of a creator or not. The ambience was noisier, and I had more difficulties to follow up all the details. A few ideas were thrown regarding the need for a creator.

At a certain point in time, Neil Armstrong made a sharp statement:

"Such a fantastic structure, with all its contents, its evolution, its rules, and the universal constants governing it, cannot be a simple coincidence."

I did not try to ask him for more details, out of respect for his private convictions. Luckily, Bisignani launched a new subject by telling Armstrong about his trip to Egypt. This was an opportunity to engage in a new discussion, comparing official history to real history.

I think it was Tim who first made a comment about the real age of the Sphinx. This triggered contradictory comments and allowed me and Tim to speak about Graham Hancock and his colleagues and their quest to correct our perception of world history. Neil Armstrong was more listening than speaking, but he made a few comments about some bizarre facts still not totally clarified by historians.

I had a few minutes to get closer to Neil Armstrong when we started leaving the table and exchanged a few words about the A380 and Emirates. He was impressed by Emirates' bold move to order a huge number of the type and questioned why the other airlines had not followed the move.

It was a pleasant and interesting interaction with such a great hero. It was also one of his last main public appearances. He died two years later.

With Tim, it was also a very enjoyable moment we spent together. Following this iconic event, we continued again our roller-coaster rides, which lasted up to the end of 2014.

For different reasons, Tim seemed not very interested in starting to look at the detailed specifications of the A350 XWB, and he was always critical of its performance.

He particularly, became furious when he heard about the change of design weights of the A350-900 and A350-1000 and made me understand that he was considering cancelling the contract. The situation became tense and was made even more complicated by the huge orders Emirates was making with Boeing for B777s.

I did my best to explain to Tim that Airbus improved tremendously the performances of the A350 and that he would have a much better aircraft than what he ordered, but he did not want to listen to me.

Luckily, during a visit to the Arabian Travel Market exhibition in Dubai, I met Sheikh Ahmed and got a few minutes to explain to him, as well as to some other executives of Emirates who were also present at the exhibition, the prevailing situation in order to be sure that the message reached all the concerned people.

When he learnt about this, Tim was not happy. He called me to express his strong anger, stating that this was a stab in his back and that he would never forget or forgive.

Although I could understand that he was not pleased, I was really surprised by the tonality of his comments.

The result of this crazy situation, which lasted for more than three years, was the order of 50 A380s in the autumn of 2013 and the cancellation of the 70 A350 contract in June 2014. It was the same manoeuvre as in 2010, cancel and order. This time, the A350 had been *killed* within Emirates.

We reached the record-breaking number of 140 A380s ordered by Emirates, but we made the way wide open for the Boeing 787 and all models of Boeing 777 to enter Emirates.

Luckily, all our disagreements between Tim and myself did not affect our true friendship, which is the most important thing I cherish in our interaction. I hate to see business conflicts destroy long-lasting friendly relations.

After my retirement, I had the opportunity to see Tim several times, and he explained to me the reasons behind some events and some decisions.

But if I summarise, all that we have done together since 1988, I can proudly but humbly say that we contributed to shaping modern aviation, each one from his own position, but with a very close and frank coordination.

Because of the vision of the Dubai leadership and under the chairmanship of Sheikh Ahmed, Emirates succeeded in innovating, developing, and growing in an impressive way, to become "the" leading airline in the world. This was achieved thanks to the founding team and, for the past 25 years, thanks to the dedication, the vista, the competence, and the strong leadership of Tim Clark.

Emirates and Tim Clark showed the way to many other airlines and caused a tremendous change in the airline landscape, for the benefit of the passengers and of their countries.

Most of my achievements, during my years at Airbus, were linked to the UAE, to Dubai, to Sheikh Ahmed and, in particular, to Tim Clark.

Sir Tim Clark has been decorated by both Queen Elizabeth and the French president of the Republic. He is Knight of the British Empire (KBE) and Officier de la Legion d'Honneur of France. He is also Fellow of the British Aeronautical Society (FRAeS).

Thank you, Sir Tim, for your friendship, which has survived all the hardships, and thank you for all that you have done for the airline industry and for aviation in general.

5.5 The Arab Spring (Won't Get Fooled Again- The WHO)

One of the most difficult periods in my life was the first quarter of 2011. The Tunisian uprising or *Revolution* took place as of the beginning of January (in reality, it started on 17 December), culminating with the departure of President Ben Ali on 14 January.

The continuous killing of demonstrators, for several days, generated a lot of anguish and stress, because it was random, and nobody was immune. In addition, we had no visibility about the future of the country, and improvisation seemed to be the master of events (later it was discovered that not everything was improvised).

Being abroad, I was frustrated not to be able to follow the action closely from within Tunisia, but some circumstances would give me the opportunity to eventually do so.

During the hot days of the revolution, the temporary building housing the nucleus of the planned factory to be established in Tunisia by Aerolia (later Stelia, now part of Airbus as Airbus Atlantic), an industrial subsidiary controlled 100% by Airbus, was stormed by demonstrators. There was no damage, and nobody was hurt, but a French citizen who was part of the task force in charge of implementing the factory died from a heart attack.

This triggered harsh decisions from Airbus, like closing the factory and repatriating all the expatriates until further notice. This was fully understandable, but there were voices within Airbus which started requesting a definitive closure of the factory and its transfer to another country (Morocco).

This factory was part of an offset obligation signed, back in 2009, by Airbus in the frame of the purchase by Tunis Air of 16 aircraft. I was not involved in the Tunis Air negotiations, but I felt sorry for the country to lose such a great opportunity to develop an aerospace industry because of the behaviour of violent demonstrators. I found myself obliged to intervene with the Airbus top management to see what could be done to change the closure decision.

Luckily, all the top executives of EADS, including those of Airbus, were present at Chantilly (north of Paris) for the yearly top executive team conference. Thanks to the great help of my friend Pierre Bayle, the—at the time—SVP Communications of EADS, I succeeded convincing Louis Gallois to issue a press release, expressing support for Tunisia and confidence in the youth of the country.

The press release left a door open for a continuation of the activities of the group in Tunisia.

As a result of the event in Tunisia, it was decided to allow me to oversee the activities of Airbus in the country. This led to a lengthy and complex battle to obtain, on one side, the support and the guarantees from the Tunisian authorities to ensure the security of the site and of the people working there, and to explain that the factory is solely owned by Airbus, with no involvement of people of the old regime; and on the other side, to convince the Airbus management to pursue the project as planned.

There were many developments after 2011, with further ups and downs, but I can say that jointly with the help of the Tunisian management of the factory and the strong support of some colleagues in Airbus and EADS, we succeeded to keep the factory in Tunisia running, and even to extend its activities. Today, the Meghira Aerospace Park is one of the pillars of the Airbus industrial system, and its activities were extended, following additional investments.

In this battle to keep the aerospace industry alive in Tunisia, there was a very strong support from some Tunisian colleagues working for major industry leaders like Mehdi Jomaa, who was the CEO of Hutchinson (a French company manufacturing aerospace components from rubber and its derivatives and which is a subsidiary of Total Group) and became later Minister of Industry and Prime Minister of Tunisia.

Here again, teamwork was key to success.

Just after Tunisia, uprisings took place in Egypt, Libya, and several other Arab countries. The situation of the airlines of the concerned countries became fragile and even dangerous, like in Libya, where some aircraft were destroyed during fierce fights around and inside the airport.

With the sales directors involved in the concerned countries, we created a sort of *crisis task force* to follow the situation and provide help when and where necessary and possible.

Two major issues became priorities: above all, send experts to assess the extent of the damages to the aircraft hit during the fights (we had to dodge the multiple failed cease-fires and the difficulties to fly to Libya), and find solutions to save the remaining aircraft; then, later, to manage the increasing debts of the different concerned airlines, mainly due to shrinking revenues.

In addition, after a few months, all the concerned airlines found themselves confronted with the need to define recovery plans to cope with the new situation.

In front of this unique situation and based on our past experience with Kuwait Airways, I proposed to both Airbus and EADS management to create a solidarity fund to help the airlines in Egypt, Tunisia, and Libya.

I was positively surprised to see EADS responding quickly and positively, thanks to the support of Marwan Lahoud, the head of SMO, and the understanding attitude of Louis Gallois, the CEO of EADS. Airbus CEO Tom

Enders also expressed his readiness to top up the amount provided by EADS, if needed.

Consequently, we tried to respond to the various demands of the concerned airlines.

I was really grateful for this gesture, which confirmed again the spirit of my company and my mother company to stand by the side of their customers when facing difficulties.

On a more personal aspect, I dedicated some of my free time to helping Tunis Air get out of the difficult situation in which it had plunged, which was the result of an accumulation of the effects of the uprising with the successive problems created over time.

I did my best, in full respect of my obligation vis-à-vis my employer, in terms of non-conflict of interest, by avoiding the fleet planning. A full audit by the ethics and compliance department was carried out, which gave me its blessing.

I was called by the Minister of Transport, who offered me the position of president of Tunis Air to restructure the company. I refused, but I suggested helping him put in place a recovery plan, working closely with the management of the airline and its president.

A solid work was done with the help of some experts and with the participation of the management of the airline. A complete report was submitted, but it was never taken into consideration. Either the new authorities of the country were not willing to move, or they were blocked by the corporatism of the employees.

In Egypt, there were less needs, and despite the drop in tourism, the airline managed to come out of the crisis with minimum damage. Nevertheless, we received sincere thanks from the Minister of Transport and the Chairman of Egyptair for our proposal for assistance.

In Libya, besides the damage assessment and the training of some pilots, we could not do more, because access to the country was difficult for several years.

During the peak of the Arab uprisings, I could count on the strong support of Pierre Bayle, a reserve colonel, an Arab speaker, and a man full of humanity, tolerance, compassion, and understanding.

Pierre had travelled a lot in the Arab world and to Lebanon, where he had been an AFP correspondent for a few years during the darkest period of the life of the country.

I was blessed when I met Pierre because he helped me a lot in removing a lot of hurdles and in communicating efficiently in the Arab world to explain the EADS and Airbus positions and to kill the false ideas carried by some media that Airbus was close to the old regimes.

The Arab Spring was a serious test for the Middle East team and for all of Airbus. We succeeded in overcoming it and our relations became stronger with all the concerned airlines.

From a political perspective, this Arab uprising against dictatorship, corruption, and very often against hard living conditions was supposed to bring democracy, transparency, and better living conditions, but it backfired by bringing to power the Islamic brotherhood that engaged in a hidden and systematic revenge against the secular states through political, social, and financial actions.

The result was the spread of terrorism, civil wars, economic crisis, and, in some cases, the destruction of the states.

How could one explain the hundreds of thousands of Arab citizens killed because of civil wars or terrorist actions?

Personally, I had a lot of hopes initially, but quickly I got a wake-up call that things were not as expected and that the set objectives of this uprising, which was not really a spring but much more a dark winter, would never be achieved. For the last 12 years, all the Arab countries concerned by this uprising saw a dramatic degradation of their stability, their security, their quality of life, and even of the little sort of democracy some of them had obtained.

It appeared very quickly that democracy, when and where it succeeded to bud, was corrupted by several factors, ranging from money to criminal blackmailing and assassination.

The populistic trend spreading around the world contributed to making the people swallow fake news and well-orchestrated propaganda and bring to power (when there were elections) persons with no vision, unable to lead clearly and efficiently, with no competence and not knowing how to save their countries from further degradation.

It was and still is a huge disappointment for me. I hope the future will unveil the truth about the Arab uprising. For me, it was also another hurdle in my life that I managed to overcome.

6. Citizen of the World (We are the World-USA for Africa)

Working in aviation, on the airline, manufacturing, or service side, implies a lot of travelling and continuous change of environment, with the inevitable consequence of becoming a citizen of the world, as you get the impression of feeling at home everywhere.

There are many positive aspects of being a citizen of the world; tolerance, acceptance of diversity, ability to adapt to variable circumstances, and easy communication with others. But there are also some negative aspects: possible loss of identity and references, and difficult family life, to name just a few.

Therefore, there is a real debate about the effect of aviation on the life of the people, and particularly those who work in it and have to travel a lot.

6.1 The World of Aviation (Jump-Van Halen)

I have mentioned above, on various occasions, some of the big events which are part of aviation life, in particular, the air shows and the IATA Annual General Meetings or *AGMs*. It may seem a bit strange to those not familiar with the aviation world that such apparently glamorous events play such an important role in this business.

However, one must realise that businesses, including those involving transactions worth billions of dollars, are, in the end, done by human beings—men and women who have feelings and for whom trust in the business partner does play an important part. This is why, after having described my two successive lives—first with an airline and then with a manufacturer—I now want to talk a bit more about what I call the *world of aviation* and what it encompasses.

Aviation is, by its very nature, truly international. Like other businesses, it may be regulated by national authorities, particularly those overseeing national air transport. But because airlines fly internationally, they must abide by all the rules and regulations that are implemented in the foreign countries they are flying to. The same is valid for the manufacturers, whose market is the whole world and not just their national carriers.

Because of this, there is—maybe more than any other sector of any industry—more coordination between all the various stakeholders to produce similar or close-to-similar standards to ease the business of flying internationally.

This is probably the reason for which there are so many transnational organisations and institutions involved, which bring together the airlines of the world as well as the manufacturers and other stakeholders, and so many international events, some of which I already referred to.

I have had several interactions with these different structures. I enjoyed some of their benefits and sometimes suffered from the consequences of their rules and regulations during my airline career.

To start with the airlines, which was my initial activity, there are many associations I came to get acquainted with, like IATA (International Air Transport Association), AFRAA (African Airlines Association), AACO (Arab Air Carriers Organisation), AEA (Association of European Airlines), ATA (Air Transport Association of America), and ICAO (International Civil Aviation Organisation).

I already mentioned SITA (Société Internationale de Télécommunication Aéronautique). I had the opportunity to represent Tunis Air on certain committees of these different associations. This enabled me to travel to many countries and meet with many airline representatives.

IATA was, for me, a forum where I met people from all over the world and learnt a lot about the functioning of the air transport ecosystem (all the committees, the clearing house, etc.) and got training sessions on planning maintenance, optimising stocks, crew utilisation, fuel consumption, etc. This helped me a lot in getting familiarised with the aviation environment I had started working in.

IATA is a unique setup in the whole world. It is an association which allows the member airlines of the world (some 80% of the world's airlines are IATA members) to work together, complement each other, and even assist each other, despite being competitors and without infringing on the antitrust laws.

First, through its different sections, it enables airlines to do interlining agreements, codeshare agreements, and fifth freedom agreements (fifth freedom is when an airline can sell tickets and pick up passengers from a foreign country while doing an intermediate stop there), sell tickets on behalf of another airline, provide technical, handling, catering, and operational assistance, share a stock of spares at certain airports, and share solutions for their common problems.

The organisation has established all the necessary tools to facilitate the life of the member airlines. The most notable is a clearing house to ensure payments between different airlines and travel agencies, like what exists between banks, with different specialised committees to manage all aspects of the airlines' activities and a sort of academy providing dedicated training courses in different fields.

IATA now also offers a unique tool to assess the safety standards of a given airline. Called IATA Operational Safety Audit (IOSA), it is internationally recognised and testifies that the airline with an IOSA stamp is safe to fly—or it makes recommendations to a given airline and helps it get up to that internationally required level.

Some 430 airlines are IOSA registered, of which the close to 300 IATA members. This means that even non-IATA members call upon IATA for such services. Safety, being the key to and basis for successful airline operations, I will come back to that aspect a bit further down.

The cooperation between airlines surprised me. I don't believe there is another industry where you can have members of the staff of a company being accepted in the offices of a competitor, to be trained, and be shown how to do the job. It is still a world of knightly competition.

But the most surprising element of this cooperation is the ability of the staff of any IATA member airline to travel on another airline for free or highly discounted (minus 90% or 75%) tickets. It is as if an employee of an oil company can get free or discounted oil gallons from a competitor. Of course, these special fares are subject to certain conditions, but they make the lives of hundreds of thousands of airline staff easier.

This was especially the case before *deregulation* was implemented in the early 80s. Before, airfares were very high and hardly affordable for a normal citizen, but after Deregulation, and especially thanks to the emergence of low-cost carriers (LCCs), fares have tremendously gone down, and nearly everybody can now afford to fly nearly everywhere in the world, and this with a guaranteed seat, which is not the case when travelling on an IATA *free* seat.

So, this early benefit is not called upon so frequently nowadays as it was when I personally enjoyed these free or discounted tickets when I was with Tunis Air (in the late 70s/early 80s), and I could travel with different airlines from Europe, from Asia, from Africa, and from the US. And, as I just mentioned, I also ended up being denied boarding and being stuck in a location for some time when flying with such air tickets.

This is what happened to me, for example, in the US when I found myself being a collateral damage of the accident of the American Airlines DC-10 on Flight 191 on 25 May 1979. I was in Scottsdale, Arizona, attending a conference organised by the American equipment and electronics company Sperry, and I had to go to New York-JFK to catch an Air France flight to Paris.

When boarding on the TWA B727, I was warned that my seat was guaranteed only to Oklahoma City, which was the next stop, and I would have to check there if I could have a seat for the second leg to JFK. Unfortunately, the aircraft was full because the grounding of all DC-10s following the accident had caused a capacity crunch and generated the transfer of many American Airlines passengers to TWA and other airlines.

Upon arrival at Oklahoma airport, the station manager called my name and the names of four other passengers (all airline staff), including my colleague from Tunis Air. He assembled us close to the cockpit and informed us that there were no more seats available for JFK. He could not tell us when we could have seats. The only good news was that my colleagues from Tunis Air and I were on top of the list and would be called first.

The station manager, who was of Palestinian origin, had heard us speak in Arabic and asked us about our origins. He then told us that there were still at least two other flights to JFK during the remaining part of the day, and all of them were arriving well before midnight, in time to allow us to catch the Air France flight.

The prospect of spending the night in Oklahoma or in New York was not much appreciated for the financial consequences, and even more so for the complications it would generate to change the bookings. Our luggage was taken out of the aircraft, and we were driven to the main building and particularly to the boarding gate dedicated to this TWA flight.

So here I am, stuck at Oklahoma City airport, waiting for a hypothetical available seat in the coming hours or days. After all the passengers had embarked, we started looking for a bar to buy some drinks. The group of five that we were became united by this misadventure, and we were competing as to who would buy the drinks.

Soon after we arrived at the bar, the station manager came to tell us that three of us (myself, my colleague from Tunis Air, and a gentleman from Air Inter, the French domestic carrier, which later merged with Air France) would be able to board on the next flight, departing three hours later. The two other persons would be able to take the flight leaving early evening.

The simple drink became a full meal, and we enjoyed a nice and friendly conversation as if we knew each other for a long time. This was the sort of solidarity we could enjoy among airline employees. We boarded as promised on the next flight, after having enjoyed a few hours at the bar of Oklahoma City airport.

On another occasion, I experienced a funny request from the airline I was supposed to fly with. It was again in the US, back in 1981. I had arrived in New York with Air France and was booked on Continental Airlines to fly to Los Angeles in first class with my wife.

When I arrived at the check-in desk, the Continental Airlines employee refused to give me a first-class ticket because I was not wearing a tie. I tried to find one in my hand luggage but without success because my wife had decided that we were to travel relaxed, in very casual clothes, and there was no need for a tie.

This was a strict rule applied at the time for all airline staff travelling in first class. It was frustrating, but I had to accept it. One is supposed to know the rules of each airline.

At that time, Continental was the subject of many criticisms concerning a supposedly sexist advertisement called *We really move our tail for you.* This was consecutive to the change of logo painted on the tail and a series of service improvements.

Since I was meeting Continental Airlines flight operations management, when I was asked how I found their new service in first class, I told them my tie story.

This led to a discussion on their advertising campaign, and one of the guys told me with a big laugh:

"Maybe you would have been accepted if you moved your tail."

This was a very good case of how a nice and innocent idea can backfire due to basic misunderstanding.

AFRAA (African Airlines Association) enabled me to discover Africa, and in particular Egypt, Ethiopia, Ivory Coast, Kenya, and Senegal. I could grasp that, despite limited resources, the other African airlines, besides Air Afrique, Air Algérie, Egyptair, Ethiopian, Libyan, Royal Air Maroc, and Tunis Air, were either very small, poor, and disorganised or still in creation.

The leading airline was Ethiopian Airlines which, despite the poverty of the country and its political instability from the mid-70s to the early 90s, was operating efficiently with a high degree of self-sufficiency.

I also saw the collapse of the airlines which were the result of alliances between several countries, like East African Airways in 1977, which was quickly replaced by Kenya Airways and by other airlines in Uganda and Tanzania. Other alliances collapsed earlier (in the 50s), like West African Airways Corporation, or later, like Air Afrique (2002).

All these alliances had been created at the time of colonisation as a tool to keep control of air transport in the ex-colonies. But this had not taken into consideration any of the new realities and the aspirations of the countries and their people, who all wanted their own national *flag* carriers.

Within AFRAA, there were a lot of opportunities for cooperation, but there was a lack of will or resources. Only a few modest projects saw the light of day, like the pool of spare parts of the B737 at some key stations. It was a real disappointment because the continent was in need of genuine cooperation between the various airlines to save on costs, but it was not the priority of the politicians who, unfortunately, controlled the airlines.

AACO (Arab Air Carriers Organisation) was a less exciting discovery because the airlines were more mature, some of them were the property of rich countries, and many had existed for several decades already. The opportunities for cooperation existed, but, as for Africa, there was no real will to do so.

Despite all the hurdles, there were a few cases, where it worked, like the joint training of pilots by using simulators of other airlines (as an example, Tunis Air did it with Air Algérie, Royal Air Maroc, and Saudia).

I was personally involved in a study for a joint leasing company, but it never materialised. This project was resurrected in 1988/1989, and I had the opportunity to participate in its materialisation, thanks to Airbus' assistance.

AEA (Association of European Airlines) contributed to my education thanks to some seminars and conferences in different fields of activity.

ATA (Air Transport Association of America) was a very important element in my training as a maintenance engineer because the aircraft, its dimensions, its weights, all its systems, how to maintain it, to manipulate it, etc., are defined as per what we call the ATA chapters (Chapter 06: Dimensions and Areas, Chapter 22: Auto Flight, Chapter 23: Communications, Chapter 34: Navigation, etc.).

ATA chapters are the basis for all aircraft manuals and for the standardisation of all the actions related to designing, specifying, manufacturing, testing, operating, and maintaining an aircraft, including the training of the personnel. I had several opportunities to attend ATA conferences, and I had to learn everything about ATA chapters in the frame of my type rating on B727/B737.

So far, for the numerous airline associations, the role of which is to defend the interests of their members at a regional and international level, and to cooperate with each other as needed. I now have to mention some others which basically regulate the air transport world and whose main objective is to ensure safety. Because, as already said, safety is key to the aviation world and the basis on which the whole air transport is built.

The most important of the international civil aviation organisations is the ICAO (International Civil Aviation Organisation). It is an agency of the United Nations which oversees all that is related to international air transportation.

As such, ICAO emits the *recommendations* that govern all aspects of an airline operation (commercial, traffic rights, alliances and fares, maintenance, flight operations including conditions for accessing airports, ground operations, procedures regarding the way an aircraft accident or major incident is handled, etc.).

Being an international organisation, it has no power to enforce these *recommendations*, but the signatory states are, however, to implement them through their respective national civil aviation authorities.

ICAO and IATA do cooperate for the benefit of air transport safety, but this does not eliminate possible tensions between the two organisations when vital interests of the airline industry are at stake. I had an opportunity to be a close witness to the evolution of certain regulations (crew fatigue, ETOPS), and I saw the diverging positions of the two organisations.

It has to be underlined that the trade unions of the different categories of airline personnel play a strong lobbying game in this, which further complicates the equation.

Then, there are, of course, the already mentioned national civil aviation authorities, which are to ensure that all the safety regulations are abided by. In the countries where airliner manufacturers are established, these national authorities also oversee the development and manufacturing of the airliners, so as to ensure that they comply with the very latest safety regulations when developing and producing an aircraft.

Contrary to airlines, the number of which exceeds 800, there are just a handful of large airliner manufacturers in the whole world. The largest two are, in alphabetical order, Airbus in Europe and Boeing (which absorbed McDonnell Douglas in January 1997) in the US, followed by Embraer of Brazil.

The Russians are also producing airplanes, but they are so far not certified in, nor do they respond to, the rest of the world's commercial and safety requirements. The Chinese COMAC, a CAAC subsidiary, has also embarked on producing an airliner, the C919, which recently started flying commercially within China.

Because the two major manufacturers are based in the US and Europe, with each also having a long history of airliner development, production, and maintenance, the civil aviation authorities of these countries have gained a wealth of experience in this respect.

The longest existing one is the American Federal Aviation Authority (FAA), followed by the European Union Aviation Safety Association (EASA), which regroups the certification authorities of the various European Union member countries. The role of these agencies is to ensure that the manufacturers, and also the respective airlines, abide by the latest safety standards, be it during the development of a new airliner and its production, as well as during its operation and maintenance.

Basically, they ensure the *airworthiness* of the aircraft and its whole operation throughout its whole life, including the way airlines operate it. These agencies also approve the airline operations, including the training standards for the flying crews. They define basic crew scheduling and rest time, etc.

They also emit *Airworthiness Directives*, the implementation of which is then mandatory by the operators. These can encompass an airliner modification or the mere reminder of proper operational procedures.

They also oversee the airport operations and all that basically is overseen by the civil aviation authorities of any country.

In fact, because of their experience and knowledge in airworthiness and safety, most national civil aviation authorities take guidance or even fully accept the lead of the FAA and/or EASA for their own standards (For example, and as described previously, the Tunisian Civil Aviation Authority had requested the French DGAC to certify their new A300 FFCC, then they did a simple transposition to certify the aircraft in the country).

It must be said that what seems to be a very cumbersome oversight, with so many agencies and entities involved, in the end ensures that air transportation is the safest of all transportation means.

This also includes all the lessons that can be learnt from thoroughly analysing in the most minute details the causes of an incident or accident, so as to learn from it in order to avoid a repeat.

It is to be noted that the objective of these technical investigations is not to apportion blame, but exclusively to *learn* from the event to avoid a repeat. The objective is merely the improvement of safety.

Because safety is considered the basis for all airliner operations, there is no competition whatsoever in this respect between the manufacturers or amongst the airlines. During numerous conferences and other such tools, experiences are shared amongst all stakeholders, so as to improve the overall safety of air transportation to the benefit of all, and above all, of the passengers and crews.

To my knowledge, there is no other industry sector—except maybe the nuclear one—where safety plays such a prominent role and with so many constructive exchanges between all members because all know that a safety hazard in the end affects them all.

I mention this because it is during such gatherings, but also the many others, that airline and manufacturer representatives as well as members of the authorities and all other stakeholders get together to exchange views and experiences. This leads to the organisation of *events* which can be very pleasant while fostering such exchanges in a relaxed—and hence more open and hence profitable—atmosphere.

I had the opportunity, during my term with Tunis Air and later with Airbus, to attend, amongst others, General Assemblies of IATA, AACO, AFRAA, and SITA. One thing was noticeable, namely the increasing involvement of aircraft and engine manufacturers through the quality of the receptions, lunches, or dinners.

The first one which impressed me was a dinner organised by Pratt & Whitney in 1978, at The Hilton Nairobi, during an AFRAA event. It was the first time in my life that I attended such a buffet! It was impressive in terms of quantity and variety of food and drinks.

It is true that, at the time, Pratt & Whitney was in a position of a quasi-monopoly when it comes to civil aircraft engines and could afford to organise such events with the objective to strengthening its relations with its existing African customers and to get introduced to the new customers (Kenya Airways, etc.).

A stupid consequence of such an event was a very serious cold caught at Orly airport when we disembarked, with my colleague Ammar Trabelsi, from the Air France B707, on remote parking, at a temperature of 4° Celsius, while wearing a simple short-sleeved shirt (we took the flight leaving Nairobi at 2 am, running straight from the dinner, without changing clothes).

These professional gatherings and events are a clear sign that despite fierce competition between the airlines, there is a sort of brotherhood spirit between them. I am still convinced that there are not so many industries where you could experience such spirit.

Here we also have to speak about how the manufacturers cooperate with each other. There are national associations of manufacturers like GIFAS in France, ADS in the UK, BDLI in Germany, and AIA in the US. There are also regional ones like ASD in Europe.

There is no real international association covering the whole world. There is one organisation called ICCAIA (International Coordinating Council of Aerospace Industries Associations), which regroups only the manufacturers of Brazil, Canada, Europe, and the US. This association is recognised by ICAO, but I don't think it has an important impact on the industry.

Unlike airlines, which gather several times a year on the occasion of the IATA AGM or the regional association's meetings, manufacturers don't gather very often during closed meetings (this might even be considered *illegal* because of the anti-trust laws). There are, nevertheless, many opportunities for the manufacturers to be present at the same locations, displaying their respective products and advertising their services and activities.

These opportunities are the airshows, which are each run every two years in sequence. There are the big ones like *Le Bourget* next to Paris (every odd year), *Farnborough* next to London (every even year), *ILA* Berlin, Dubai Airshow,

Singapore Airshow, Moscow Airshow, and many others smaller in size, in many countries around the world.

There are also dedicated air shows for some aviation segments like AERO Friedrichshafen in Germany for general aviation, NBAA in the US or EBACE in Geneva for business aviation, and Aircraft Interior Expo in Hamburg for cabin interiors, etc. There are also exhibitions dedicated to maintenance, catering, training, and all the activities related to aviation.

These airshows, which last between three days to a full week or eight days, have common features and routines.

They include hospitality lounges—also called chalets—with excellent food and beverage, under the supervision of renowned restaurants, receptions and gala dinners organised in iconic locations, multiple press conferences and media briefings, and last but not the least, the display of as many products as possible on the dedicated interior stands and on the tarmac display area, and of course, the public's favourite—the highly acclaimed flying displays.

Some airlines are also present with a *chalet* at certain airshows, but they have lower profiles than the manufacturers.

The stars of the show remain the big manufacturers through the sizes of their venues, their stands, and their open-air display areas.

An air show is the pinnacle of communication and media coverage for the manufacturers. For this reason, when possible, the signing ceremony of a big contract during airshows (even if the deal has been finalised for a while) is an event that is very much sought for.

With my successive teams, I had many signature ceremonies at different air shows, with all the razzmatazz that goes with them, from the signature itself behind closed doors to the press conference, then to the photo session, and finishing with the VIP lunch.

For the manufacturer employees involved in the air shows, it is a double-edged sword. On the one hand, it is exciting and somehow entertaining, but on the other side, it is stressful, frustrating, and exhausting to manage it all.

The daily shuttles between the hotel and the location of the air shows, are by themselves, a nightmare and a big source of tiredness and frustration due to huge traffic jams and the time spent in cars and buses to move over dozens of kilometres. And then the many late-night dinners when you would dream of a mere sandwich and an early night!

In summary, one of the most important consequences of the existence of all these events (conferences, seminars, air shows, etc.) is the possibility they offer all stakeholders in the aviation sector—from the airlines to the colleagues of the aircraft and engines manufacturers, as well as the component suppliers—to meet and to know each other better, to exchange views and experiences (especially in the field of safety, thereby greatly contributing to its improvement and making air transport the safest of all transportation means), and to discuss jointly some common problems while fostering trust. These events are really an essential tool for the aviation business overall.

In this respect, one can say that there is a large aviation community, and despite, sometimes, severe competition, there is a unifying spirit—*the love for aviation*. For that reason, those who are not sincerely passionate about aviation cannot last long in their positions.

Before closing this review, it is worth mentioning that the global landscape of air transport experienced a real earthquake on 24 October 1978, when President Jimmy Carter signed the *Airline Deregulation Act*, which allowed airlines to have the freedom of deciding the fare level they want to charge, the choice of routes and of timetables. Airlines could also decide the type of service they would provide and the airports they want to operate to.

As a result of Deregulation, beyond the iconic TWA and Pan Am, other US carriers could now also fly international and transatlantic routes, in particular to Europe. This triggered a deadly price war between them and eventually, combined with other internal factors, led to the demise of Pan Am and TWA.

This also caused a big wave of mergers and acquisitions in the US, leading to the unexpected result that eight mega airlines (American Airlines, Delta Airlines, Southwest Airlines, United Airlines, Alaska Airlines, JetBlue Airlines, Spirit Airlines, Frontier Airlines, etc.) today control 95% of the traffic—actually, the opposite of the initial objective of Deregulation!

This Deregulation wave eventually also reached Europe and the rest of the world and, there too, reshaped the air transport landscape. This was a good opportunity to launch the all-new concept of Low-Cost Carriers (LCCs), which started mushrooming. Close to 120 were created between 1978 and 1984, with a new paradigm which obliged the legacy airlines to review their strategies, leading to a big wave of mergers.

The knockoff effect of Deregulation went far down the chain of air transport: from the average passenger, who could now afford to fly, down to the airline, further down to the aircraft and engine manufacturers, and reaching ultimately the suppliers and the sub-suppliers. Within less than 20 years, which is the normal life cycle of an aircraft, we saw a complete change in the rules of the game.

Air Transport had been completely regulated, with everything predefined, including the fares, which were agreed upon on a bilateral basis at the IATA level once a year and which allowed airlines to have a relatively long-term visibility for their business, allowing, consequently, the manufacturers to plan their new products accordingly.

Post-Deregulation, the environment became totally variable and, in a way, uncertain. Every month, every week, and even every day can see changes in different parameters of the business, from fares to destination to the type of service. This was supposed to push the manufacturers to be more agile and plan according to these shorter cycles.

Unfortunately, a lot of constraints (starting with the safety requirements during the development of an airliner, the existence of a technology plateau which limits the extent of the possible improvements using today's technology, and the need to satisfy the stakeholders, particularly the shareholders) made it very difficult for the manufacturers to bringing substantial innovations.

It is true that aircraft technology saw tremendous development from the early '70s until today, but there was no real breakthrough in terms of speed and usable energy. A quantum jump is badly needed. It will be very costly and will, above all, take time.

Despite all the ups and downs, my Tunis Air experience was fantastic. I learnt everything about real aviation life in this small, modest airline based in a developing country, which was on par with the airlines of developed countries and was even a pioneer in some critical fields.

Over the years, we were dozens of Tunis Air managers, engineers, and pilots who emigrated to other countries to work for other airlines, for maintenance companies, and for manufacturers. I had very good echoes about all these experiences, and I am proud of some ex-colleagues' careers.

My experience in Tunis Air gave me also a very good lesson: aviation is a very structured field of activity, mainly based on human resources and competencies, and as long as you respect the rules and the limitations, you can do a lot of things and be successful. Financial resources play the role of an enhancer of the capabilities of the airline. If you have only the money and you don't have the people, you cannot do anything, unless you *import* the people.

Tunis Air had the people because the country, thanks to President Bourguiba, bet on education and allowed the independence generation to be as educated and competent as the citizens of the Western countries. This had ended the segregation between the Europeans and the locals that had existed, in terms of education and job opportunities, during the colonisation period.

It was not a question of intelligence or intellectual capability (an African or an Arab can do the same job as a European), it was a question of deliberate policy, to impose the leadership of the Europeans on the citizens of the colonised countries.

Aviation was a real symbol of liberation.

6.2 Aviation and Politics (Political World-Bob Dylan)

One of the key elements I discovered while working in the aviation field, either on the airline or the manufacturer's side, is how important politics are in the conduct of the companies' business.

As far as the airlines are considered, they are either privately or government-owned. Governments tend to use them as a tool for the development of the country. This is normal and acceptable as far as this utilisation fulfils the real needs of the population, helps develop businesses, and creates wealth.

Unfortunately, in most of the countries where the airline is owned by the government and the regime is not really democratic, the airline becomes a simple personal power tool in the hands of the President or ruler.

All key appointments within the airline are decided by the leader of the country and are very often made on the basis of loyalty, and in several cases, it concerns members of his *extended family*. Competency in the field of aviation and business running is regrettably not the top priority.

This type of situation led, in many cases, to mismanagement and other abuses, which caused the quasi-collapse of such airlines. But they were often saved by injection of public money because the leader likes to keep his tool (or toy) in his hands for as long as possible.

Another important element in the paradox of survival of quasi-bankrupt national airlines in many countries is the fact that the leadership knows that if the national airline disappears, this will cause big discontent among the people. Since this option is not acceptable, cash injection becomes a systematic remedy.

Effectively, and despite the abuses, national airlines always play a vital role in the economy of the country, and they have quite often a dual role: serving the country and serving the leadership.

In some countries, where there is freedom for the labour forces to be organised under the banners of trade unions, there is the possibility of a junction between the leadership and the trade union to keep the status quo at the airline. Each side likes to protect its interests. The leadership can use the airline as it wishes, and the union keeps the jobs and all the other advantages forever, reaching even some type of hereditary transmission of the jobs.

The combination of dictatorship and corporatism is deadly for the airlines, which must be a multi-role player. I really admire such airlines to continue operating in such conditions.

But the link between an airline and the political leadership can also have very positive effects on the country and on the airline. Let us take the case of the landlocked countries. Without an airline, they cannot have any link with the rest of the world. I admire the success of Ethiopian Airlines, which is the best airline in Africa from all aspects, despite the difficulties of the country.

It is an airline which operates according to high international standards, with a very modern fleet, makes profits, and above all, serves the country well.

I followed Ethiopian for more than 40 years, and it has been always on top, even during the darkest period of the country, during the rule of the DERG (the military committee ruling the country following the *coup* against Haile Selassie, the former emperor of Ethiopia), between 1974 and 1987.

The resilience of such an airline and its capacity to adapt to variable situations should be taught in aeronautical institutes.

A country like Rwanda put a lot of effort into setting up and developing a national airline to become the tool of its development. It opens the horizons (it is a landlocked country) and acts as the vector of exchanges with the rest of the world.

Here is a case where the intervention of the state in the airline business was well managed, with the right approach, and was even necessary. There are more cases like this one around the world.

For many other countries, like Hong Kong, Qatar, Singapore, the UAE, and others, the airline became the national symbol and the key tool for their expansion and for the hub concept. All the above-mentioned countries are not big in size, but their airlines are amongst the biggest in the world, supported by very modern and huge airports, catering for tens of millions of passengers per year.

All these countries are world hubs, and their airlines are part of their economic strategies. Here again, there is a quasi-confusion between the airline and the state, but these airlines are managed according to modern and efficient standards and are usually profitable. And when not, as in one case, the benefits it brings to the country overall still largely compensate for its deficit.

Let us not forget that aviation is one of the first expressions of the independence of a country. During the first half of the 20th century, aviation was under full control of the US and Europe. Japan, which had had a prominent role before WWII, lost it after 1945.

The name of the game was *No white, no flight*. When countries in Asia, Africa, and elsewhere started gaining their independence after WWII, the first thing they nationalised was the airline that was previously controlled by foreign interests linked to the occupying country.

It took around a decade for the newly independent countries to train their nationals to become pilots, technicians, operations, and commercial staff to run the national airline. The main objective was to be, as soon as possible and as much as possible, independent in all the fields of activity.

The pride was to have all the cockpit and cabin crews, as well as the maintenance and engineering personnel, being nationals and to do the maximum of aircraft maintenance in the country. For the political leadership, the airline became a strong expression of independence.

Regarding manufacturers, the role of politicians is as important as for airlines, with multiform connections, whether the manufacturer is private or state-owned.

First, most of the manufacturers have a military component in their business and need government support for funding and for selling domestically and internationally. The sale to the national military establishment is vital for any manufacturer. That is why one could say that the relation between a manufacturer and its country's Ministry of Defence is somehow entangled. This has been the case since decades and is valid worldwide.

On another aspect, the manufacturers need, for both their civil and military businesses, to export. To do so, the diplomatic support of the governments is very often necessary. In some cases, even a certain bargaining power is required. I have given some examples earlier on in this book.

But politicians also like to use the manufacturers to improve their image. How many times did we see presidents, kings, prime ministers, or ministers of Airbus home countries keen to be present at contract signatures, new aircraft rollouts, inaugurals of factories, and airshows?

On some occasions, and in the early '90s, I was confronted with politicians who tried to convince me to appoint lobbyists to promote our sales in some countries. On one occasion, a high-ranked advisor of a prominent politician went as far as making threats to *convince* me.

Luckily, Jean Pierson, to whom I reported the problem, approved my refusal and protected me. From that date on, I refused to meet any politician alone. I managed always to have a witness. You never know!

The two sides need each other, whether people like it or not. On the one hand, aviation is glossy and glamorous, and on the other, politics is attractive and tempting. There is a magnet effect which creates a bridge between the two.

Yet, this bridge can also be positive and beneficial for both sides. I could observe many cases whereby government officials and ministers were appointed CEOs or chairmen of airlines or aviation manufacturers; and the reverse, whereby top executives of airlines and aviation companies became ministers and even prime ministers.

I remember that Royal Jordanian saw one of its presidents become a prime minister (Nader Dhahabi) and its VP of Communications become Minister of Tourism (Akel Baltagi). Both succeeded in their respective missions. Both were good friends and were competent and dedicated people.

The same happened in many other countries all over the world, including in Africa, Asia, Europe, and the US, with some failed experiences, but globally, things went relatively well.

6.3 Aviation Leaders (Airlines' Leaders) (The Boss-James Brown)

Thanks to my cumulative experiences in the airline sector and in the aircraft manufacturing industry, I had the opportunity to observe the human qualities, the social behaviour, the leadership abilities, the competencies, and the performances of dozens of airline leaders and executives.

There are a few key conclusions which could be drawn from these observations, which spanned more than 40 years.

First, there is a great disparity in the profiles. It ranges from low-educated persons with strong political backing to highly educated ones (PhD and above). It could be somebody from the ranks of the airline who climbed the ladder to an appointed person from outside the airline (often a former politician or high-ranked civil servant).

He or she (there are more and more female airline leaders) could be an engineer, a pilot, a financier, a business manager, etc. The spectrum is large, but since the '90s, we started seeing a new trend of professionalisation of the airline leadership, where a group of professional CEOs COOs, CFOs, etc., started moving from one airline to another, like football players moving from one club to another. This trend is mainly seen among the Western airlines, some Asian carriers, and the leading airlines of the Gulf region.

These airlines are not inhibited by any nationalistic consideration and don't have a problem to appoint a foreigner for the sake of improving their operations and financial situation. The recent move of Air France to hire Ben Smith, a Canadian coming from Air Canada, was a good illustration of this trend.

At the same time, there is a large group of airlines, mainly in the developing countries, where the airline is part of the symbols of sovereignty and appointing a foreigner is not admitted. I must confess that there are some arguments in favour of this nationalistic approach. Since the '50s, most of the countries succeeded in building an aviation infrastructure and training their citizens in the various fields of aviation, and in particular the airline industry.

Therefore, the probability of finding the right person for the right job is high, but the political and personal considerations often lead to the wrong choices. Some airlines, like Ethiopian, succeeded in making the right choices, while others failed. In addition, like in football, the salary of a local CEO or COO is very often lower than that of an expatriate.

Second, a substantial portion of airline leaders are extremely cultivated and could easily speak about a wide range of subjects. The rest were really focused on aviation but nevertheless had, in their majority, a good knowledge about the geography and the political life of the countries to where their airline operates. This was one of the key elements in my decisions to develop or not close relations with the airline leadership.

It was much more interesting and rewarding to get close to people from whom I could learn new things than to have a one-way relationship with somebody from whom you don't hear anything new or of interest. I had cordial relations with all the airlines' leaders, but some were much closer to me.

Unfortunately, culture seems not to be correlated with the level of education. I found many PhDs not being very cultivated, while others, with a much lower lével of education, were a shining and inspiring source of knowledge in many fields.

But there is one field where I found common interests with a large part of the airlines' leadership, and that is music. Whether it is Arabic, Oriental, or Western music, many chairmen, CEOs, and executives were passionate about music, and I was able to share opinions about songs and even sing with them on some occasions. I have vivid memories of one evening in Rio de Janeiro, returning from the location of an IATA event, where, for some reason, half the bus was singing 'Hey Jude.'

These moments of unison were very rare and were full of joy and friendship. Race, nationality, language, company, hierarchy, etc., were ignored, and the sole interest was to sing in tune and in full harmony with the others. Through the gala dinners and the musical shows displayed during the airlines' gatherings, music became a sort of additional unifying element.

Third, even if many aviation leaders are highly educated and very cultivated, some of them nevertheless develop strong egos, which occasionally lead to some quite funny situations. Again, it has nothing to do with good human qualities.

I have some CEOs and other executives of many airlines who are good friends, with whom I enjoy spending time and engaging in discussions, but when you watch them closely in their daily lives and during events and visits, you can detect this strong ego. I noticed the same with some executives of Airbus. It was really shocking in some circumstances.

I'll never forget somebody who refused to get in an Audi A8 because it was not a Mercedes S-class, as he had requested. A chairman of an airline refused to take a bus with other CEOs and top executives to attend a special event organised by the host airline during an IATA general assembly. He arranged his own transportation with a luxurious car. He arrived very late to the event because his driver had gotten lost a few miles before reaching the location of the event.

The road was complicated because it crossed a small forest, and it was not on Google Maps or any other navigation system. I would never have suspected this sort of behaviour from this person because he looked humble and easygoing. But this was not the case. I eventually discovered his real nature and his big ego. I tried to have a logical explanation for this big ego story, but I could not find anything robust.

It is true that the majority of those holding top positions in society—whichever their field of activity—tend to develop big egos. Airlines executives are, therefore, no different from others.

They are exposed to a lot of pressures from different sources, and they enjoy some privileges (like travelling for free, often in first class, staying in top-class hotels, and being in close contact with the political leadership, especially in the developing countries where there is a national airline). I believe the combination of all these factors makes them develop a sort of shield (consciously or unconsciously).

I really hated being exposed to these egos because, even if they try not to express it directly, many airline leaders could not control themselves in some stressful conditions, and then their real nature appeared through their acts. This was one of the main causes of my disagreements with some airline executives.

Due to the nature of my job, I had to swallow a lot of unpleasant comments or actions, but I never tolerated anything which could hurt my dignity and/or my reputation or the dignity and/or reputation of my colleagues and company. Each time, I reacted strongly but politely. In some cases, it even put an end to my relationship with the concerned person. This was the price I accepted to pay, and I don't regret it.

Fourth, airline leaders have one of the shortest durations in their positions compared to other industries. This is mainly valid for the developing countries. It is true that during the last decade, things seemed to have been improving, with key executives staying longer in their positions, probably because they now tend to be appointed more frequently on the basis of their competencies.

This situation was justified by many factors: the frequent wrong choices, the political factor (appointing cronies), the difficulties to improve the situation when the leader has a limited margin for manoeuvre, the blocking role of some trade unions, which favour corporatism over the interest of the airline, and finally, the inherent challenges of the air transport industry.

Many airlines woke up to the danger of the frequent changes of management and started giving more time to the key position holder to analyse the situation, define the solutions, implement them, and correct them as needed. In the case of three airlines I followed closely, the average durations of the CEO were, respectively, nine months, one year and a half, and two years. None of these three airlines was profitable for several years.

But to be fair with airlines, there were, and there still are, many exceptional leaders who succeeded to put their airlines on the route to success and even in the leading positions. History will remember their names. I refrain from listing them in order not to upset those whom I could forget.

In summary, I'd like to emphasise that, no matter the egos of some, all the persons working in the aviation world are all exceptional and, above all, passionate about what they are doing for that industry. Because, I believe, the fact of having to work internationally broadens everyone's perspective and view of the world.

You must learn and understand the other and their point of view. You must learn about other people's cultures and ways of thinking and doing things. This is one of the key elements which, to me, makes this world so fascinating (as said above/below in the following chapter).

Other aviation leaders.

I also had the privilege to meet several leaders of manufacturers and leasing companies. They are as diverse as the airlines' leaders, but in their gross majority, they played a big role in the shaping of the actual aviation world.

The list is long, but going back in time, I had the chance to meet people like Thornton(T) Wilson from Boeing, Jack Welsh and John Rowe from GE, George David from United Technologies, Sir Dick Evans from BAe, Jürgen Schrempp from Daimler-Benz (DASA), Steve Udvar-Házy from ex ILFC (which he created in 1973), or Olivier Andriès from Safran.

All of them impressed me with diverse degrees, and I could, with a few of them, have lengthy discussions, which allowed me to know them better and appreciate their qualities. With Steve Házy and Olivier Andriès, I developed real friendly relations which never stopped.

One of the most interesting encounters I had was with Sir Dick Evans. Jean Pierson called me to the Airbus in-house *Le Club* restaurant, where we used to invite our guests, to join him during his lunch with Sir Dick Evans. They were busy eating their heavy lunch, accompanied by some good French wines, and I had to brief them on the progress of the Saudia campaign, giving some details which required their full attention.

At a certain point, while I was speaking, they started exchanging views about the quality of the wine and comparing it to another one. I thought I'd need to repeat my explanation, but I was surprised to hear Sir Dick Evans asking me a precise question about an important detail mentioned in my presentation.

Jean Pierson noticed my surprise and said to me in a teasing tone:

"You think we are not able to follow your explanations while we appreciate the quality of the wine? Well, we wait for your answer."

I spent close to half an hour with them. They offered me to share the cheese platter with them, but I had already had lunch, and I was so impressed by these two guys that it was difficult for me to join them for that. I did not have either their power or their appetite.

All these leaders played a role in making aviation what it is today, and all of them deserve respect and gratitude. Some of them were—or still are—pillars of this industry. One special mention to my friend Steve Udvar Házy, who made the leasing business one of the key elements of the air transport industry, by helping airlines find other ways to acquire aircraft than through a straight purchase.

A simple look at the percentage of leased aircraft, out of the total number of aircraft operated worldwide shows that Steve won the bet he made when, in a bold move, he started his leasing company (International Lease & Finance Company, or ILFC) in 1973—and by far.

I met Steve for the first time in 1977, a few months after I joined the Tunis Air technical department, as a member of a committee negotiating the lease of a B737-200 from ILFC. I had two challenging but interesting meetings with him, during which I discovered the personage and the leasing business.

I learnt about his Hungarian origins, his birth in Budapest, his immigration at the age of 12 with his parents to the USA, transiting through Sweden, his studies at UCLA, and his beginnings in business. I was impressed by all that he had already done at the age of 31.

At that time, ILFC was a small company, and its portfolio was comparatively modest. The management team was very small in size but highly competent, and there was a strong desire to answer customers' requirements promptly and efficiently, with Steve being on every front. But progressively, he put in place a very competent and efficient management team, with John Pluger as his loyal number two.

I lost sight of Steve for a few years before meeting him again after I joined Airbus. We had several common projects for leasing, and we succeeded to place many ILFC aircraft in the MENA region as well as in other countries during my time as Senior Vice President of Sales worldwide.

De facto, I was, chronologically, the first to meet Steve among all the Airbus executives, and we remain friends until now.

Steve Házy is one of the few key players who, like Tim Clark, reshaped air transport and aviation in general, through his multiple inputs in aircraft design and through bringing new solutions in sourcing aircraft. I admire his energy and his joyful attitude.

He continues to be an important player in the leasing business through his new company, ALC.

6.4 World Leaders (World Leaders-Rem)

In the frame of my activities in either selling passenger aircraft or corporate jet aircraft, I had the opportunity to meet several heads of state, prime ministers, ministers, and princes (some are mentioned earlier in this book). I am not going to relate the details of my encounters with all of these personalities. Nevertheless, there are some common features which deserve to be explained and analysed.

All of them showed a passion for aviation and did not hesitate to engage in very detailed technical discussions about aircraft specifications and performance. Whatever their background, all of them were keen to learn more about aircraft and dared to ask all sorts of questions.

With many princes, the interaction was very simple, straightforward, and full of mutual respect, but with one or two of them, it was complicated, with a clear desire by the concerned person to impose a relationship of superiority. This was mostly the case of princes with business activities.

The main problem with all these leaders was up to what detail we could go into when speaking about aviation. It was easy to use precise aeronautical expressions with King Hussein, President Mubarak, Prince Bandar of Saudi Arabia, or Sheikh Mohamed Bin Rasheed Al Maktoum. All of them were or are pilots.

The most interesting discussions took place with Prince Bandar, who explained to us some tricky flights during his training, and with King Hussein, who explained some manoeuvres in a very detailed manner, as only pilots could do. I remember when he described a stormy arrival at Biggin Hill with a Tristar in night condition.

He was the co-pilot, with the captain being Miss Taghreed Akacha, one of the first—if not the first—female pilots in the Arab world. Taghreed is a Palestinian from Jerusalem, and she is a strong character.

King Hussein explained how he had a disagreement with *his captain* on the approach procedure and how he accepted to follow the order of the captain, adding:

"There is only one captain on board."

This was a clear illustration of the humility and professionalism of late King Hussein. He was a real airman, and he was not flying for the fun of it, but out of passion.

One evident conclusion is that many of the world leaders I had the chance to meet in my life played a very important role in shaping the aviation of their respective countries, particularly in the developing world. I will never forget the visionary speech of Sheikh Mohamed Bin Rashid Al Maktoum, the leader of Dubai, during my first meeting with him.

The far-sighted vision of some of them transformed the air transport industry beyond the borders of their own countries.

6.5 Where is Home (Sweet Home Alabama-Lynyrd Skynyrd)

I used to go for running with Nick Tomasetti, the former CEO of Airbus North America and previously a prominent executive of Pratt & Whitney, whenever we met in any city around the world. One day, during a Farnborough Airshow, we were jogging in Hide Park late in the afternoon in July when we suffered a sudden drop in temperature and a strong headwind pushing us back, making us freeze.

Without any prior communication between us, we started laughing and screaming:

"No way, it is again the same thing."

The other colleagues who were running with us stopped and asked us what was happening. Nick explained that we had a similar situation in Tampa, Florida, at the Saddlebrook resort in January 1985, when, during our daily run, we suffered the same drop in temperature and the same freezing wind.

It is true that January 1985 saw a wave of very low temperatures all over the Western Hemisphere, from minus 18° in Toulouse to minus 9° in London and Atlanta, and to minus 5° in Tampa, where IAE (International Aero Engines—a joint venture between Pratt & Whitney, Rolls Royce, MTU Aero Engines of Germany, and Japanese Aero Engines Corporation (JAEC)) was organising a seminar to present the recent improvements introduced on its engine which was proposed for the A320.

IAE was not lucky with its seminars, because the next one, organised in Newport, suffered a violent hurricane.

After returning to the hotel, we sat down with Nick and a few other colleagues for a drink and engaged in a philosophical discussion. What is home? The triggering point was that all of us were travelling a lot, had gotten used to the same places in different cities and countries, faced similar situations in different locations, and ended up living the life of nomads.

There was a serious debate following a comment made by Nick, who said:

"I need an anchor to which I can relate and use as a reference, and say: this is my home."

But one of the colleagues asked him:

"Is it your actual home in Washington, or the previous one in East Hartford, or is it your birthplace?"

In the end, all of us agreed that the home is the one you relate most, both emotionally and physically, and not simply the one you are living in with your family.

This discussion is still in my mind, and I asked myself: what is really my home? Without hesitation, I concluded that it is my house in Carthage, Tunisia. It's the place where I have my roots, where I can connect with the members of my family, and where I can feel the history and the heritage of my country. It does not mean that I don't enjoy my life and my home in Dubai, or I don't enjoy spending a few days in Paris or Toulouse.

The big difference is that in Carthage, nobody can tell me that I am an alien. In Paris or Toulouse, despite my French passport, I am always labelled as being of Tunisian origin. In Dubai, I am an expatriate resident, even if I am well integrated into the life of the city and the country, with a lot of very close friends among the UAE nationals.

I still rotate between France, Tunisia, and the UAE, enjoying every place, but on the day of my departure from this life, I know my ultimate place will be Carthage, Tunisia.

Many of my colleagues faced serious crises when trying to accommodate the different constraints generated by the need to remain close to their country of birth while they lived in the country where they were working and spent a lot of time in different locations scattered all over the world. It is a big challenge for them.

Many could not cope with it, and they ended up having to break up their families or take hard decisions—like ignoring their country of origin or, in some cases, sacrificing their professional careers.

This is one of the few shortfalls when working in some sectors of the aviation field, especially when working in *commercial:* being on the move all the time and not being able to relate to a reference which makes you know who you are and where you come from. But, on the other hand, those working in aviation and staying there despite some of these shortfalls do it because they are passionate about this industry and are happy to make their career in this sector.

As said in the case of Steve Házy, you can meet someone at an early stage of your career when working in one sector and meet that person again several years later when in another company. In the end, those working in aviation make up for a real *family*—a family of aviation enthusiasts and passionate individuals.

This may be the most striking and significant characteristic of aviation. You feel at home in the aviation community, which makes up a kind of small village in which you meet people time and again as they move on!

6.6 The Global Village (In the Name of Love-U2)

The notion of the global village was brought in by several experts, philosophers and sociologists among whom we could mention Marshall McLuhan, to describe the new world generated by globalisation which is itself a consequence of the ease of transportation, and the ease of communication and access to the media, from any place in the world, thanks to the dazzling progress in the communication and information technology.

This globalisation was further extended by the rapid standardisation of food, drinks, entertainment, clothes and several other aspects of daily life around the globe. As I said, before, besides the language and sometimes the features of the people, a modern traveller, is not anymore able to know, easily, where he is.

I am not going to comment on the pros and cons of globalisation; I leave it to the thousands of experts who are working on this touchy subject. I am just interested in the role of aviation and how it transformed the world over the past 70 years, or since the dawn of the jet age.

By allowing people to move fast from one place to another at reasonable prices, aviation became an accessible commodity. It encouraged people to leave their villages, their cities, their regions, and their countries for education, for research, for work, for business, for war, for love, for leisure, and even for crime.

The result of this ease of move is that you can find any citizen or any product from any country in any other country, and you can witness activities, customs, religious events, or event celebrations of a community of one country in other countries. (Chinese New Year or Indian Diwali are celebrated in many countries; Halloween is everywhere, etc.) One could say that aviation is the real link builder between the citizens of the world, for its good and for its bad aspects.

I should add that air transport also facilitates the exchange of many goods around the world—for good or bad. But one thing is for sure: without the air transport industry, the perishable goods such as vegetables grown in winter, say in Africa, could not be made available in more Nordic countries. More than anything, by doing so, air transportation unquestionably contributes to the economy of developing countries.

As mentioned before, air transport was able to build a unified community between the different airline staff through dedicated structures and tools and thanks to very innovative agreements, which are unique to the industry. As already mentioned, it is the only industry where an employee of one company can have access to the product of another company free of charge or at a largely discounted price.

I don't think an employee of McDonald's can have a free lunch at KFC, or an employee of Sheraton can have a free stay at a Sofitel or Kempinski hotel. This community-building role of air transport expanded to encompass the whole world population.

The COVID-19 pandemic (2020–2022) and environmental concerns were seen by many experts as game-changers which would tremendously reduce the role of air transport in favour of the internet (online work, virtual meetings in lieu of physical meetings, video conferencing).

Personally, I was convinced that the airline industry would face a sharp drop in the volume of business, even after the end of the pandemic, which I forecasted effectively for mid-2023 during a videoconference in May 2020. But here, I was wrong! It is true that the virtual office has become very popular, videoconferences are very frequent, and internet shopping is a daily routine.

But the human being needs, at least at a certain point in time, a physical contact in his dealings with real people. In addition, he or she is always keen to travel for leisure and for family considerations.

This explains the rebound in air travel demand and the difficulties faced by airlines to cope with this unexpected—but most welcomed—rebound in 2023.

This means that persons will continue to travel by air and will continue to do it frequently. This induces further integration of the global village. But what will the governing rules of this global village be? This is the key element for defining the future of globalisation.

Are we going to have a single country? Or a block of countries imposing directly or indirectly their social lifestyle, their values, their culture, their economy, their products, and their services to the rest of the world? The village would be uniform and not as attractive.

Or are we going to have a simultaneous coexistence of different social lifestyles, values, cultures, economies, services, and products? The village would then be multiform.

But there is another possibility, which consists of having the different countries and blocks of countries agreeing on common governing rules, allowing diversity in all aspects of life, with all countries and communities having a say in the build-up and in the running of the global village.

The role of the airline industry will depend on the option chosen or imposed. The real restructuring of air transport will take place after the emergence of the shape the global village will take in the next 20 years.

But another element will also define the future of air transport. It will be the respect of the environmental constraints, which will dictate the type of energy to be used and the limitations for the use of flying vehicles.

7. Lessons Learnt
(I Believe I Can Fly)

The only thing I learnt from aviation, is that my knowledge is limited. I know a few basic elements, but I am ignorant in many areas. I sincerely admire the experts who are called, often on TV or who speak on social media, to answer questions about aviation events and who dare to speak about any subject, ranging from manufacturing an aircraft to operating it, to maintaining it, or servicing it, but without any real insider knowledge.

As far as my modest experience could allow me, I can say that aviation is a fundamental pillar of the world economy and is playing a unique role in bringing the citizens of the world together and contributing to their economy. Therefore, it is here to stay as long as mankind exists.

7.1 Aviation is a Real WWW (Lift Me Up-Moby)

One main lesson I learnt is that aviation constitutes a real worldwide web, as efficient and as useful as the internet. It connects people, it allows exchanges of goods, generates wealth, and employs millions of people all over the world.

Very often, we ignore all the activities around aviation and supported by aviation. Some figures computed a few years ago suggested that if the aviation ecosystem were a country, it would be ranked among the 20 wealthiest and most productive ones in the world, even if the margins are minimal.

This mere fact is important to consider when it comes to targeting aviation for pollution or to impose additional taxes. All the governments are happy to see aviation generating wealth and employment (while airlines are hardly profitable), but accuse the same airlines of many sins and tend to make their lives difficult.

It is a pity to see how different the attitudes are towards the internet and aviation while the two are providing similarly valuable services. During the last pandemic, the two W.W.W were even opposed to each other: internet is safe and eco-friendly—without taking into consideration the exact effect of the huge energy consumption by the data processing centres of the different service providers, simply because they are not visible—while aviation is dangerous and pollutes.

We saw some people asking to take advantage of the situation generated by the pandemic to reduce forever the role of aviation. But on the other side, everybody was expecting to swiftly get vaccines, which could only be brought in a timely manner by this very same aviation—something the internet is unable to do!

We all agree that it is vital to protect the environment, which is a common asset for the whole of humanity. But, as just seen, the role of aviation is vital and cannot be the subject of erratic decisions dictated by passion, political, or philosophical considerations.

I see a strong contradiction between the push for the global village and the rising criticism vis-à-vis aviation.

We must have a coherent approach towards the future of humanity and not act with scattered, contradictory decisions.

7.2 Keep on Flying (Keep on Running-Spencer Davis Group)

Despite the criticism of environmental activists regarding the contribution of aviation to the pollution of the planet, I don't see a substitute for an airplane to transport people and goods from one point to another safely, quickly, and at an affordable price during the next 100 years. Consequently, man will have to continue to fly, and aviation will remain a pillar of the world economy for decades to come.

There is a strong possibility to reduce the role of aircraft overland by developing high-speed trains in the coming 20 years. In Europe, North America, the Gulf region, and in big countries like China, India, or Russia, the role of aviation could be reduced, but it would need huge investments both in infrastructure and equipment, mainly in vehicles.

This explains that, up to now, most of the countries elected to invest in aviation (airports and aircraft) because the total cost is very often lower than the rail option—and may be even more ecologically friendly!

As one rightly observed:

"Four kilometres of runway give you access to the world, but four kilometres of railway lead you nowhere. Think of all the land needed for a sustainable railway system!"

Furthermore, an efficient railway system requires an efficient electrical power supply system. But to get electricity, you must produce it. Unfortunately, until now, apart from a limited number of countries (like France), which have an important part of their electricity generated by nuclear plants, most of the world runs on fossil energy.

Despite the risks linked to the storage of the waste of nuclear plants, the atom remains an acceptable bargain until humanity becomes able to produce enough power from renewable or other more ecological energies. One century ago, we were not able to produce energy from atoms. Decades ago, we were able to control properly nuclear fission and produce energy.

We started also controlling nuclear fusion to produce the H-bomb, but we are still not able to produce efficient energy. Fusion is the future because the raw materials needed for the nuclear reaction are available in huge quantities. The energy produced is colossal, and the waste is clean and safe.

Even if we envisage the fission option, the chances for humanity to find some safe and efficient technical solutions to treat waste during the next two centuries are very high, while the expected life of the waste is counted in thousands of years. On the other hand, the option to continue with fossil energy will lead to a very dangerous pollution of the environment within decades, with catastrophic consequences.

When we see the slow pace of deployment of renewable energies, we can feel the urgency of making some harsh decisions. Time is at a premium. Aviation, despite its very modest share in the pollution of the atmosphere, is one of the most exposed and needs to act swiftly.

Man will continue to want and need to fly. It is a commodity which will remain forever—unless humans will be able to use tele-transportation, like in *Star Trek*.

8. The Future (In the Year 2525-Zager & Evans)

Nobody can predict what will eventually happen, but anyone who follows closely aviation could have some ideas regarding the different possible options for the future. As already mentioned, we are all aware of the environmental and energy challenges, and the continuous and growing simultaneous demands for speed, comfort, and lower costs.

Airlines are facing a dilemma because, with the current level of technology, they cannot respond to all three requirements. That was the reason, as explained earlier, for the segmentation of the market between low-cost airlines, charter airlines, and legacy carriers, with an additional segmentation between short-haul, medium–haul, and long-haul, as well as limited-service and full-service. Consequently, in order to draw any view about the future, we need first to make some basic assumptions.

The first assumption is about speed. Do we consider supersonic flights to become the rule or not? The debate is open, but higher speed will prevail.

The second assumption is about hubs or no hubs (flying point-to-point and avoiding the hubs). The debate is still open, but point-to-point will prevail, together with the existence of some major hubs.

The third assumption is about the available energy. From what I observe today, and based on the different running projects to build new aircraft powered by renewable energy, I would guess that the most likely option is a hybrid aircraft powered by a dual power train using electricity and hydrogen, with the first step being electricity and biofuel (like the VoltAero Cassio family of aircraft).

When the supply, storage, and utilisation of hydrogen become really easy and safe—which at the present stage still represents a significant technological challenge—the thermal engine will be replaced by a hydrogen-powered engine (the basic design of the engine is similar). It is then possible to envisage using only hydrogen to power aircraft.

Some people think that a nuclear-powered aircraft could be envisaged. I don't disagree with that idea, but this solution could only come much later when a lot of stringent requirements are met.

In any case, these three assumptions will define the shape of aviation in the future.

There is another element which will shape the future of aviation: multinationalism. The majority of the countries of the world will have a contribution to aviation, either through local technical capabilities, through investment in big international companies, or through the competencies of their citizens working locally or internationally.

A century ago, only a given number of European countries and the US were able to build aircraft and, later, space vehicles. Today, even many developing countries are able to build aircraft (including my country, Tunisia, with its rising star, Avionav), and some of them are able to build space vehicles and send men to space.

Fifteen years ago, Airbus and Boeing were afraid of the potential competition of the Bombardier CSeries aircraft, with a capacity between 100 and 150 seats. Then, in late 2017, Airbus bought the aircraft, which was then renamed A220.

Both manufacturers thought they were back to the comfortable duopoly. But a new competitor is emerging in the Far East—China—with the Comac C919. If we add the Brazilian Embraer EC190, there are four single-aisle jet aircraft, with more than 100 seats, available on the market.

In close to a century, the world went from *No white, no flight* to *Anyone can be an airman*. This is another proof that aviation was—and continues to be—a tool for emancipation and development.

What the Gulf and Asian airlines have been doing for air transportation, and what China and other Asian countries have been doing for the aerospace industry during the last 40 years, is huge compared to what Europe and the US have done over the last 120 years, considering the respective periods of time.

The aerospace centre of gravity is, slowly but surely, shifting from the West to the East, with all the consequences that implies for the future.

The continuous rise of the Global South (with the BRICS being the main driver) is surely affecting the future of aviation in the long run.

The chances to see the new generation of big aircraft powered by renewable energy taking shape in the Global South are greater than to see it emerging in the West. The environment in some countries, where the intervention of the state is still a normal practice, will favour courageous decisions which will help make breakthroughs in many fields, contributing to the advancement of aerospace technologies.

In the so-called West, some red tape and the disproportional importance of the shareholder value (Let us make more money before investing in new technologies) will hinder a lot of courageous decisions.

Just observing the huge steps made by China in the field of building powerful and efficient batteries gives an idea of the possible future for electrically powered aircraft.

I believe Europe has to take courageous decisions now to avoid being progressively sidelined or left behind in shaping the aviation of the future.

The Global South will not wait for the West to set the pace.

Conclusion
(Before Landing)
(Rien De Rien-Edith Piaf)

I had, so far, an exciting flight of life, full of everything, from the best to the worst.

It was not a smooth flight, it was much more a bumpy ride, flying through heavy storms, with also some very smooth rides. But overall, the positive outcomes outweigh the negative ones.

From a simple passion of a kid and later of a teenager, aviation became my job. Against the belief of many people, I learnt its theoretical basis in Toulouse, France, and its practical basis in Tunis. I owe my career to the quality of education I was granted, both in Tunisia and in France.

The years at Tunis Air, the modest airline of a North African country, were very useful in shaping my technical and managerial skills and in making me able to work anywhere in the world.

Airbus gave me the opportunity to use my acquired expertise and express my skills worldwide.

I had the chance to participate in some pioneering projects which contributed to shaping the aviation of today, both with Airbus and Tunis Air.

For some mysterious reasons, it seems that my fate was to have my life linked with Airbus from the day I met Bernard Ziegler in 1976, while I was still a student.

I had also the chance to be part of some important events which left their fingerprints on our recent history.

My life and my career were influenced by some key persons, to whom I'll continue to express my gratitude until the end of my life.

It has not been always easy, but I can say that I represent the new human type—a citizen of the world, living in a global village.

I don't believe my flight was perfect, but nobody could doubt my will to do my best and my desire to demonstrate that a citizen of a developing country can be as competent and as efficient (sometimes even better) as a colleague from Europe and the US. This is not the result of a feeling of inferiority, but much more of a feeling of injustice.

I read a lot of books written by the dominating powers about the countries they dominated. I saw how truth was bent and how facts were distorted to prove the superiority of the dominant power. The best examples are the books written by Roman historians about Carthage, after its criminal destruction, and by French writers about North Africa to justify colonisation.

I am a strong admirer of Hannibal, our national hero, and I always follow his motto: *EITHER WE FIND A WAY OR WE'LL BUILD IT.*

This is the reason why I never give up, even at my present age.

Playlist

It is a sort of a sound track of the book which could help the understanding and the apprehension of the different situations described by the author.

1- Hit the road Jack (RAY CHARLES)).
2- Awaken (YES).
3- Johnny B. Goode (CHUCK BERRY).
4- Breathless (JERRY LEE LEWIS).
5- The End (THE DOORS).
6- The Monster (STEPPENWOLF).
7- Beautiful Noise (NEIL DIAMOND).
8- Rocketman (ELTON JOHN).
9- Teenage Rampage (SWEET).
10- Airport (THE MOTORS).
11- Hold your hand (THE BEATLES).
12- In the army now (STATUS QUO).
13- Walking on the moon (THE POLICE).
14- The Letter - Give me a ticket for an aeroplane (THE BOX TOPS).
15- O Toulouse (CLAUDE NOUGARO).
16- Speed of Sound (COLDPLAY).
17- Stairway to heaven (LED ZEPPELIN).
18- Paris s'éveille (JACQUES DUTRONC).
19- Sweet dreams (EURYTHMICS)
20- Communication breakdown (LED ZEPPELIN).
21- Atom heart mother (PINK FLOYD)
22- Crossroads (ERIC CLAPTON)
23- Interstellar overdrive (PINK FLOYD).

24- Get back to where you once belonged (THE BEATLES).

25- Night in Tunisia (DIZZY GILLESPIE).

26- 4 & 20 years ago (STEPHEN STILLS).

27- Sultans of swing (DIRE STRAITS).

28- Satisfaction (THE ROLLING STONES).

29- Front line (STEVIE WONDER).

30- Bohemian rhapsody (QUEEN).

31- Under pressure (QUEEN/ DAVID BOWIE).

32- On the road again (CANNED HEAT).

33- Simply the best (TINA TURNER).

34- Magic carpet ride (STEPPENWOLF).

35- Take a chance on me (ABBA).

36- Don't you forget about me (SIMPLE MINDS).

37- Space oddity (DAVID BOWIE).

38- Englishman in New York (STING).

39- Won't Get fooled again (THE WHO).

40- We are the world (USA FOR AFRICA).

41- Jump (VAN HALEN).

42- Political World (BOB FYLAN).

43- The boss (JAMES BROWN).

44- World leaders (R.E.M).

45- Sweet home Alabama (LYNYRD SKYNYRD).

46- In the name of love (U2).

47- I believe I can fly (R. KELLY).

48- Lift me up (MOBY).

49- Keep on running (SPENCER DAVIES GROUP).

50- In the year 2525 (ZAGER & EVANS).

51- Rien De rien (EDITH PIAF).